DELTA TEACHER DEVELOPMENT SERIES
Series editors Mike Burghall and Lindsay Clandfield

Going Mobile

Teaching with hand-held devices

Nicky Hockly and Gavin Dudeney

DELTA
PUBLISHING

Published by
DELTA PUBLISHING
Quince Cottage
Hoe Lane
Peaslake
Surrey GU5 9SW
England

www.deltapublishing.co.uk

© Delta Publishing 2014

ISBN 978-1-909783-06-5

Edited by Mike Burghall and Lindsay Clandfield
Designed by Christine Cox
Cover photo © iStockphoto.com/derreck
Page 78 photo © iStockphoto.com/Devonyu
Page 105 photo (bottom right) © Ladislav Faigl

Printed in China by RR Donnelley

Acknowledgements

The authors would like to thank the many language learners and teachers with whom we have trialled activities in this book. Their feedback has helped us identify challenges and choices in the use of hand-held devices, and how best to exploit them in the language classroom.

We would especially like to thank the teachers who appear in our case study videos: Carla Arena in Brazil, Julie Cartwright and Lisa Besso in Australia, Alicia Artusi in Argentina, Helen Collins in Spain and Jon Parnham in Hong Kong.

And special thanks go to the teachers and students of the Anglo European School of English in Bournemouth, UK, for allowing us to share their excellent student projects as examples of the kinds of things you can do when you let your students work with mobile devices.

Finally, *Going Mobile* has benefited from the excellent guiding hands of editors Mike Burghall and Lindsay Clandfield, as well as the design prowess of Christine Cox. A heart-felt thank you to all of you for your hard work and support throughout the writing of this book.

From the authors

I've never been much of a gadget person.

Ask me what make of car so-and-so drives, and I'll shrug and say 'a blue one'. I've always been immune to the 'shiny box' syndrome – that propensity to fall in love with glossy new gadgets.

And although I love using technology, I am hard-pressed to say exactly when it was I started using computers or what programs I cut my word processing teeth on. A true geek would not only know – but be able to reel off – the exact specifications of their first Amstrad computer.

But I do remember my first mobile phone. And as mobile devices have got better at doing a whole range of things – from making phone calls, to showing me new places to eat out; from taking photos, to telling me how to get from A to B (and in the accent of my choice) – I have started to appreciate them more and more.

However, this is less for their looks than for the central role they play in my life and, increasingly, in my teaching and training.

For several years now, students have been bringing mobile devices like phones and tablets into their language classes, but many teachers are as yet unsure about how to deal with this relatively new phenomenon. It is still unusual to find teachers who actively encourage their students to use mobile devices in the classroom, in creative and innovative ways.

There are still many teachers out there who need help in getting to grips with the new mobile technologies – not so much in the 'technical' sense, as in the pedagogical sense: how to integrate these devices into their teaching, in a principled and pedagogically sound manner.

This book aims to provide that help.

Grounded in our own experience of teaching and training with mobile devices, Gavin and I offer you a range of activities and ideas – as well as a thorough discussion of the challenges, choices and considerations – of going mobile.

We hope you enjoy *Going Mobile*.

Nicky

I've always been a gadget man.

Or – more precisely – I've always been an early adopter, buying and playing with new technologies and working out what they can do for me, and what they might mean in the context of my work. So much so that, at home, I have several large metal boxes full of old gadgets, dating back to the early nineties, which I somehow can't bring myself to throw away.

To me, these gadgets are a visible reminder of how much the world has changed in the past twenty years.

Just about ten years ago (way back in 2003!) I went to a talk at an IATEFL conference about the value of texting in ELT – the first talk on 'mobile learning' I ever saw.

Fast forward: through countless talks about simple mobile devices, the publication of David Crystal's book on 'txtng', the British Council's early experiments with mobile learning … and on to today – where mobile phones and tablet computers are everywhere and always 'on'.

But not in our classrooms, where they are still the elephant (albeit a small one) in the room.

These days, we hear talk of 'disruptive technologies' – and the unexpected developments in mobile and hand-held devices over the past few years must be one of the most disruptive innovations we've seen for quite some time.

From basic phones to smartphones, the mobile device is the first connected technology that has managed to get into the hands of a significant proportion of the world's population.

However, it is one which, to this day, is largely ignored in education.

So Nicky and I are delighted to get this book into your hands.

There is so much potential in these small mobile devices, and we hope you will take this opportunity to get them back out, turn them back on …

And take them into class.

G.D.

Contents

From the authors   Page 3

Part A   Page 7

The big questions   Page 8

The big issues   Page 15

The big challenges   Page 24

Types of apps   Page 29

Part B   Page 31

Chapter One   Page 32
Hands off!

Features and functions   Page 33

Mobile me   Page 34

Technology timelines   Page 34

The perfect phone   Page 36

Addicted!   Page 36

Cold turkey   Page 37

Safe and sound   Page 37

Don't do it!   Page 38

Mobile classroom   Page 39

Mobile rules!   Page 40

Chapter Two   Page 41
Hands on: Text

Know your letters   Page 42

Know your numbers   Page 44

Sticky boards   Page 45

In the clouds   Page 46

Heroes   Page 47

Networks   Page 48

Mr. Ed   Page 49

Twitter celebrities   Page 50

Short and sweet   Page 51

Txtng   Page 52

Face off   Page 53

Very flash   Page 54

All in the mind   Page 55

Chapter Three   Page 56
Hands on: Image

Got it!   Page 57

Close-ups   Page 58

Find it!   Page 59

Word bank   Page 60

My other me   Page 61

Picture puzzle   Page 62

Time will tell   Page 63

Photo fit   Page 64

Contents

All around me	Page 65
Picture this	Page 66
The name of the game	Page 67
In the 'hood	Page 68

Chapter Four **Hands on: Audio**	Page 69

Musical me	Page 70
Ringtone movie	Page 71
Daily noises	Page 72
Inventions	Page 73
Listen to me!	Page 74
Lost in translation	Page 75
The place I love	Page 76
This is my life	Page 77
What is it?	Page 78
Talking trash	Page 79
Crazy sports	Page 80
Follow me!	Page 81

Chapter Five **Hands on: Video**	Page 82

Silent movies	Page 83
On the vine	Page 83
Visual poems	Page 84
Bollywood	Page 85
It's mine!	Page 86

A life in film	Page 86
It's news to me	Page 87
Come with me	Page 88
This is my song	Page 89
Show and tell	Page 90
Trail blazer	Page 91
Book bonanza	Page 92

Part C	Page 93

Going further in your classes	Page 94
QR codes	Page 97
Tactile screens	Page 100
Geolocation	Page 102
Augmented reality	Page 106
Going further in your institution	Page 110
Looking back …	Page 117
Looking forward …	Page 118

From the editors	Page 119
From the publisher	Page 120

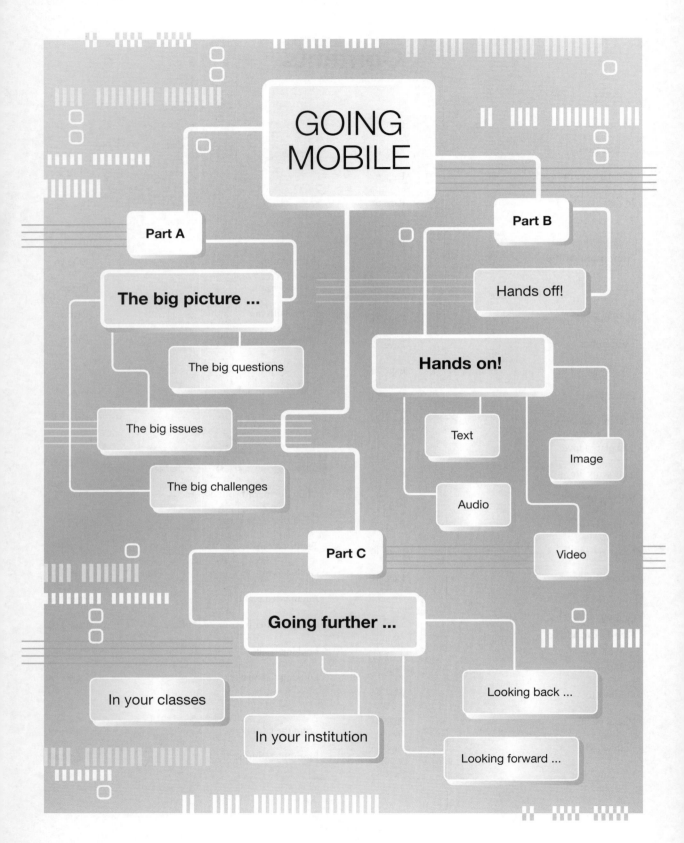

GOING MOBILE

Part A

The big picture ...

The big questions

The big issues

The big challenges

Part B

Hands off!

Hands on!

Text

Image

Audio

Video

Part C

Going further ...

In your classes

In your institution

Looking back ...

Looking forward ...

Going mobile

There is something truly inspiring and almost revolutionary about these scenarios:

- You are watching a group of learners take out their mobile phones – and really put them to good use in the classroom.
- You are sitting down with your learners – as they edit the video materials they gathered using tablet computers on an excursion into town.
- You are listening to pairs of learners talking to each other – about their favourite photographs on their mobile phones.

It's the feeling you get, as a teacher, when an activity so engrosses a group of students that they almost forget they are talking in English, sitting in a language class.

In many ways, the love of the *tool* can pave the way for an increased commitment to – and interest in – learning the *language*. And if learners can undertake part of their study time with their own choice of tool, the interest is often higher, or more sustained.

Mobile learning

Of course, 'mobile' doesn't just refer to the tools themselves, but to the opportunity for study outside the classroom – on the move. 'Learning on the go' is one of the educational buzzwords of this decade, a follow-on from distance and online learning – something truly in the hands of the learners, available anywhere and at any time.

Mobile and hand-held learning is perhaps the first technology-based approach that has escaped from the fanciful dreams and desires of educators – to the realm of the possible, and even desirable, amongst learners.

Whether it's on a mobile phone, a hand-held tablet computer or other portable gadget, more and more people are taking advantage of the technologies both inside and outside the classroom to extend their learning and find real opportunities to put it into practice. In *Going Mobile*:

- We look at what mobile and hand-held learning is.
- We explain how you can get started with it.
- We demonstrate how you can set about fully ensuring its principled and effective implementation in your own context.

So let's get going.

In Part A, we look at the 'big picture' of mobile and hand-held learning, and consider some of the major questions, issues and options – as well as looking at some sample case studies from around the world – before we investigate the kinds of challenges you are likely to meet, as you experiment with mobile learning in your own teaching or training.

The big questions

This overview will help you in your initial preparation, before you move on to try some of the activities in Part B. Our aim is to prepare the ground for your own experimentation and reflection, and to answer some of the key questions you may have about mobile learning.

What is mobile learning?

Whilst mobile and hand-held learning are considered relatively new additions to the teacher's armoury, the use of mobile phones, at the very least, has a history dating back over a decade now – a period of time that qualifies it as an old or 'embedded' technology in teaching terms! As long ago as 2003, Hamish Norbrook (then of the BBC) – writing in the Guardian newspaper[1] – was extolling the virtues of the text message in class:

'Text messages offer opportunities for the English teacher because they provide a realistic basis for writing exercises. Fewer and fewer letters are being written – especially informal ones. Yet writing is as relevant as ever – and increasingly the ability to recognise a variety of different registers, formal or informal, serious or light, is essential. Often the exercise need not be a long one. So, instead of saying "Write a letter", say "Write a text". With predictive texting the language can be as formal as you like – and although current text messages have space for only 160 characters, with a multimedia messaging service (MMS) you can write far more.'

It has, however, taken a lot longer for most educators to notice the power of these mobile devices, and it often seems as if we have moved on very little from that period, with many people still looking on mobile and hand-held devices as primarily text-based platforms.

The truth, of course, is much more complex than this, and today's mobile devices (primarily smartphones and tablets) incorporate a set of features which make them ideal for classroom use, and which replace a slew of tools that we have been using for many years: the audio recorder and the video recorder being just two examples.

It is high time, then, that we re-evaluate these devices for their potential use in class.

Yet few people working in the field can agree on exactly what constitutes mobile learning, and definitions (Kulkuska-Hulme 2009[2]; Traxler 2009[3]) are notoriously hard to find – as 'mobility' itself is a problematic concept, depending on where the focus of the word resides. When we talk about mobility:

- Are we talking about the mobility of the learners (in the 'anytime, anywhere' model of blended or online learning)?
- Are we referring to the devices themselves, and their portability?
- Are we referring to the context in which the learning takes place – in formal classroom settings or informally, elsewhere (Sharples et al 2009[4]; Kulkuska-Hulme et al 2009[5])?

These are the issues that are current in discussions of mobile and hand-held learning, and which are now beginning to be considered and addressed in research and in a new wave of publications in the area.

Mark Pegrum (2014)[6] divides these different areas into three categories for language learning, each with a different emphasis:

1 Learning that takes place when the **devices** are mobile.

2 Learning that takes place when the **learners** are mobile.

3 Learning that takes place when the **learning** is mobile.

The first category places the focus firmly on the *device* itself, not on the location or situation. In this mobile learning scenario, the learners will typically be engaging with content on mobile and hand-held devices, perhaps accessing online resources or creating them, but will be in fixed locations such as a classroom, study room or even home.

In this sense, they are not physically mobile, nor are they necessarily taking advantage of some of the more revolutionary features of mobile devices, such as geolocation or similar ('geolocation' refers to a mobile device's ability to know where the user is geographically located at any given moment, and can be used by the device to deliver relevant information to the user, based on that location).

Thus the *learning* is not 'mobile', and corresponds more closely to more traditional CALL-based activities and approaches.

The second category envisages *learners* on the move, as they work with – or create – content. This may be moving round limited spaces, such as the classroom or institution, but may also encapsulate other spaces such as the home, the bus on the daily commute or similar.

In this case, they may be working with discrete content such as flashcards or grammar exercises from major publishers and organisations, catching up with reading materials or listening to podcasts.
For example:
- Cambridge University Press – *http://www.cambridgeapps.org/*
- The British Council – *http://learnenglish.britishcouncil.org/en/apps*

In any of the above cases, the *learning* is similar to more traditional approaches, the only difference being that the learners themselves are on the move.

Pegrum's third category envisages a tighter *integration* between what happens inside and outside the classroom, and a stronger link between learning content and experiences and learning opportunities outside the classroom. Learners will have opportunities to work with real-world content, to incorporate parts of their normal lives and to interact with the environment around them.

In this kind of *learning*, they may use a tool to record new vocabulary as they travel around their city, subsequently using the tool for listening practice as they revisit each place.
For example:
- Woices – *http://woices.com/*

These are useful divisions – and ones that we feel allow for a comfortable development cycle for teachers wishing to experiment with integrating mobile and hand-held learning into their current practice.

We have based the activities in Part B partially on this model – allowing for an on-going development in terms of complexity, and with a clear progression within each chapter towards the kinds of activities envisaged in the third category of Pegrum's taxonomy.

Why is mobile learning important?

Few technologies have embedded themselves so firmly in the lives of people as the mobile phone. Indeed, mobile phones are almost everywhere you look, all around the world, and they are one of the everyday objects that most people aspire to own.

A simple look at per capita mobile phone penetration as far back as January 2012 (a survey by IndexMundi, from CIA World Factbook data[7]) shows 100 countries worldwide with greater than 100% mobile phone penetration, and many more countries below that index with a significant spread of mobile technologies.

There are many reasons for this, of course, but the principal ones are largely economic – from the cost of handsets to the end user, to the sums involved in extending the infrastructure throughout whole countries.

This is especially true of larger countries, or countries with complex geographies:
- Mobile phone signals reach further than other communications technologies.
- The devices needed to receive them cost less than many others.

Whilst you are highly likely to find more traditional technologies – such as desktop and laptop computers – in much of the world, you are increasingly likely to encounter countries and tiers of societies which have skipped this stage of technology development and moved straight to mobile and hand-held devices, for the reasons we have described.

And it is, as we note below, not simply a question of mobile phones, but of mobile *devices*: devices such as tablet computers and other tools that we can carry with us, and which provide us with some kind of link to the world beyond our regular spaces.

Also, these mobile devices have achieved something where most other technologies have failed, and that is to become ubiquitous, ever-present and almost invisible to the end user. In many ways, they have achieved a degree of normalisation that Bax (2003)[8] could only have dreamed of at the time.

Where Interactive Whiteboards (IWBs), computer labs, laptops and other technologies have only made relatively small inroads into education on a global scale, mobile devices are already present in a lot of our classes all around the world, and are ready and waiting to be exploited by both teacher and learner.

Mobile learning, then, *is* important:
- It can take advantage of the most common global technology.
- It can help learners understand and appreciate the power of the everyday technologies they carry with them.
- It can open up a path between what happens in the classroom and what happens outside, in the real world.

It is also, in some ways, a great leveller – more of which below.

Who is mobile learning for?

The short answer to this question is that it's for everybody, everywhere – wherever there are sufficient devices to go round, and where there is interest from both the teacher and the learners to try it out.

The longer answer is a much more varied picture, in which many factors need to be evaluated and balanced out – to ensure that mobile learning implementation can be carried out successfully and with the involvement of the widest possible number of stakeholders.

There are also wider issues to consider with *any* technologies in education:
- Learner attitudes to technologies in their education.
- Teacher attitudes to technologies in teaching.
- Institutional regulations.
- Technical support.
- Infrastructure: hardware, internet access …

We will look more closely at some of the major challenges later in Part A of *Going Mobile*, and in greater depth at institutional implementation/planning in Part C.

What equipment is needed?

Although we will largely be concentrating on mobile phones and tablets, mobile learning can come in a variety of packages:
- From the humble 'dumb phone', through to feature phones and smartphones.
- From e-readers to games consoles.
- And a lot more!

Let's take a look at a few of these common devices:

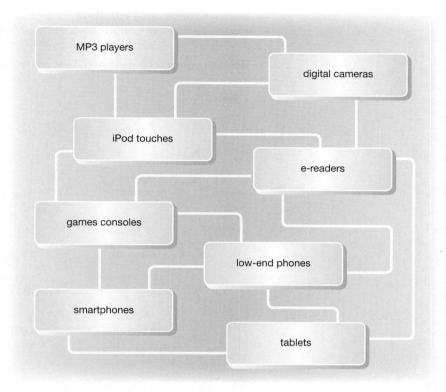

Many of the devices in the diagram can be used for language practice 'on the go':

- **MP3 players** can give the learners access – through tools such as podcasts – to authentic listening material in an area of interest to them.
- **Digital cameras** can provide the learners with opportunities to take photos as prompts for producing language.
- **iPod touches** (essentially smartphones without the phone part) can run apps and connect to the internet.
- **E-readers** can provide meaningful reading opportunities, from novels to blogs, newspapers and magazines.
- **Games consoles** can also provide language practice, both from the language of the game itself to possibilities for connected chat with other players, or through word games and other language-related puzzles.

All these mobile devices provide opportunities, mainly, for learners to *consume* language in one form or another.

If, however, you are more interested in having learners *produce* language, then some of the other devices in our diagram may be more applicable:

- Even **low-end mobile phones** can capture data for language practice – notes, text messages, photographs and more.
- More modern devices such as **smartphones** and **tablets** package a wide variety of tools and features (or 'affordances' – as we will be referring to them) into one portable tool. The inbuilt camera is an excellent tool for capturing 'language in use' and working with that language back in the classroom – as is the audio recording function. And the note taking function can provide a rudimentary way of recording new language.

This 'convergence' is a feature of modern technologies, and one that works very much in the teacher's favour, combining, as it does, a set of tools which would have necessitated a suitcase just a decade ago.

Of course, it is rarely a simple question of one approach or another, and a solid combination of opportunities to practise and produce language can be put together with good use of a variety of the tools and devices we have featured.

There are also a variety of peripherals and add-ons for smartphones and tablets, making them even more useful and versatile. Some of these include the elements listed below:

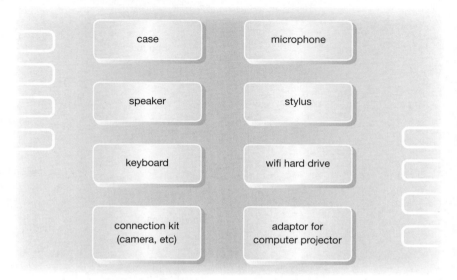

- **Cases**: These are vital to protect phones and tablets, especially if they are being used in schools. Good solid shockproof cases are available in a variety of styles, colours and budgets. Some even include the following:
- **Keyboards**: If you plan to have your learners producing a lot of written text, then take a look at some of the Bluetooth keyboards available for both smartphones and tablet computers.
- **Styli**: A stylus can help with navigating a device, but will really come into its own for note taking, sketching and mindmapping activities.
- **Microphones/speakers**: Sound-based activities can be enhanced considerably with better-quality audio input and output devices. These are not strictly necessary, but a good microphone gives you more opportunities to capture audio and work with it at a later date.

One last hardware consideration is what you, as the teacher, will do with *your* device, in terms of modelling usage and sharing materials. In short, how will you share *your* device screen with your *learners*?

There are basically three methods of doing this:

- **Hardware-based projection**
  This will typically involve plugging an adaptor cable into your device and connecting it directly to a projector or IWB system (note, however, that this will not give you IWB-like functionality from the device). Whilst this is one of the more robust ways of projecting your screen, it does leave you effectively tethered to the desk where the cable reaches, and – perhaps more importantly – adaptor cables have a bad habit of falling out of many tablets and smartphones.
- **Wireless-based projection**
  If you have good wireless connectivity where you work, you may be able to take advantage of devices such as the Apple TV to 'mirror' (duplicate) your device's screen to a screen or a projector via the wifi network. Although this is also essentially a hardware-based solution, the advantage is that you can move freely around, as long as you are within range of the wifi signal.

■ **Software-based projection**
This method routes the audio and video signal from a mobile device through a piece of software on a computer connected to a projector. As this is also linked to wifi strength, you'll need a good wireless internet setup at work, but will be free to roam. For example:
■ Air Server – *http://www.airserver.com/*
■ Reflector – *http://www.airsquirrels.com/reflector/*

Both of these two projection options have the advantage of allowing the learners to also project *their* screens (with appropriate permissions or passwords) and thus enable a greater degree of sharing and collaboration.

What content is available?

As with most teaching scenarios, there is a choice between making your own materials (or getting the learners to produce content) or using pre-packaged content from providers such as publishers, app developers and other teachers.

In terms of producing content, modern phones and tablets take the idea of 'convergence' and package it in small and easy-to-use devices, making them ideal content and language production tools. As we will see in some of the activities in Part B of *Going Mobile*, much creative work can be done with these devices in and out of the classroom without installing any further apps or spending any further money, or even without any connectivity, although their potential expands exponentially when they are connected to the wider world and the larger 'conversation'.

Apps, however, can bring new functionality to a device, from podcasting to blogging, from video editing to comic strip production – and beyond. To date, much of what has been produced specifically for the ELT market is not overly transformative, nor does it take full advantage of the convergence aspect of many devices, but there is still a very solid selection of practice apps for learners – from most of the major publishers and organisations such as the British Council, down to smaller, niche developers.

To get started with this kind of content, you may want to explore the following links:
■ Cambridge University Press – *http://www.cambridgeapps.org/*
■ Oxford University Press – *http://elt.oup.com/searchresults?q=apps*
■ Macmillan – *http://www.macmillaneducationapps.com/*
■ The British Council – *http://learnenglish.britishcouncil.org/en/apps*

Looking beyond these ELT apps, which tend to focus largely on testing and practice rather than on production, there is quite a choice available – from the Apple App Store, through Google Play and, increasingly, for Windows phones. Currently, these are really the only three key players in the mobile device arena, particularly in terms of versatility and app availability.

As with most other essentially online resources, the challenge for most educators is knowing what's out there, and where to find it. Here are some suggestions:

■ Sign up for free app sites.
For example: *http://appgratis.com/gb*
Apps 'go free' regularly, and it's worth keeping an eye on this kind of site to see what is free each day. Some websites will allow you to create an account and set up a 'watch list' of apps you're interested in – if and when they become free, you will receive a notification. If you're interested in ELT apps from providers such as the ones listed above, it's worth 'liking' them on Facebook, as they will often announce special offers there.

■ Follow mobile learning blogs and review sites.
For example: *http://www.scoop.it/t/elt-apps*
Sites such as these will provide useful links to – and often short reviews of – apps and related ideas connected with language learning and mobile devices. In this case, the ScoopIt site has regularly-updated links to a variety of resources connected to mobile learning and ELT.

- Organise an 'app share' at your next local teachers' event.
  An app share involves people getting together at a conference or event, and telling each other about the apps they like and use regularly. This can be done informally in small groups, or with everyone being allotted a small amount of time to talk to the whole group.

One major advantage apps have over software for larger platforms such as laptops and desktops is that they are generally much cheaper to buy, and can be installed on a greater number of devices at the same time.

- This is particularly true of Apple devices such as the iPhone and iPad, where one account can be synchronised to a number of devices.
- For larger-scale projects, there exists a Volume Purchase Programme (VPP) for education, allowing you to buy a number of copies of a given app at a reduced price.

As with regular desktop software, there are also often 'light/lite' or free versions of apps, allowing you to sample them and test their appropriacy and usefulness – before committing to deployment with any particular group.

How this evaluation is carried out will depend mostly on your context and situation, but a general sample evaluation rubric such as the one below may give you some ideas for your own strategy:

Evaluation table	
Name	Strip Designer
Version	V2.0
Developer	Vivid Apps – *http://www.mexircus.com/Strip_Designer/*
Platforms	iOS
Price	$2.99
In-app purchases	None
Description	- *This app allows you to create comic strips based on your own photos, either from the camera roll, or by taking them directly with the camera.* - *There are different layouts and over 100 templates.* - *You can also draw on the photos, and add speech bubbles and more visual effects.*
Input options	Camera, Camera Roll
Output options	Camera Roll, Email, PDF, Dropbox, Flickr, Facebook, Twitter
Potential uses	- *Dialogue writing, short scenes, social language, joke telling, and more.* - *Traditional dialogue-building activities.* - *Good for collaborative writing projects, literature and reader summaries …*

Kathy Schrock[9] provides a useful breakdown of apps in various areas, mapped to Bloom's Revised Taxonomy – the classification of learning objectives within education, proposed in 1956 by a committee chaired by Benjamin Bloom and revised in 2000.

The big issues

These, then, have been some answers to the major questions concerning mobile learning, and we can now consider some of the major issues and challenges posed by 'going mobile'.

The models of implementation

There are various ways of implementing mobile learning and, in this section, we will be taking a look at the most popular. You will find more in-depth information about institutional implementation plans in Part C, but here we give a general overview of the principal decisions and considerations involved.

Class sets

A typical class set implementation will have one, or several, sets of devices available for teachers to borrow and use on an ad-hoc basis. An academic director, librarian or a technical support person or technology learning officer will usually manage class sets centrally. Teachers will sign up for a particular class, collect the devices and return them at the end of each session.

In our first case study, the teacher shares a wide range of impressions and suggestions connected to a class set approach, showing how all decisions are interrelated – as they are in any advance into new territory.

Case study 1: Class sets in Brazil

Teacher: Carla Arena
Video: *http://youtu.be/rjU24EUqkCo*

Carla's school uses class sets of iPads with several groups of learners. The main aim of the project is to enhance language learning through a student-centred approach to the use of the devices, with the devices being used for communicative language production tasks.

Carla highlights the importance of training teachers, and allowing them to become comfortable with the devices before using them with learners. She also emphasises the importance of having reliable wifi connectivity.

Carla's advice to teachers is:
- First focus on the aims of your project, and the logistics.
- Involve your team of teachers in the project.
- Take an mLearning online training course.

In many ways, being in control of the devices makes for a more comfortable experience for the teacher. With a carefully managed class set of devices, these can all look and work the same, have the same apps installed and – with judicious use of 'cloud storage' (see below) and email accounts – communicate with each other and share resources easily.

The devices can be charged overnight, stored in a safe place, protected with solid covers and – where necessary – be devoid of distractions such as Facebook accounts or Twitter feeds. This kind of approach is ideal for institutions where mobile learning is not a mandatory and daily part of the teaching, but where the resources available can be shared around the classes on an as-needed basis.

For the teacher, class sets have many advantages for the people managing them and implementing them, but they may present considerable challenge and frustration for the *learners*, who will often need to learn a new system, find their way around what is essentially someone else's device, and work with choices that have been made for them.

These learners will typically have little or no choice as to what is installed on the devices, and little say in how they are used. Unless provision is made for the learners to take the devices outside the class space (either to other locations, or home, for further use), learning is generally firmly anchored in the physical space of the classroom and much of the advantage of being physically mobile is lost.

Case study 2: Class sets in Australia

Teacher: Julie Cartwright
Video: *http://youtu.be/w9nCi_Uatl4*

Julie is a teacher who uses a class set of iPads with her learners. She encourages her learners to record videos, take photos or search the internet with their devices. For example, she has used the devices in a lesson on modal verbs of deduction, in which the learners take close-up photos.

She highlights the importance of having reliable wifi connectivity, and also describes the challenge of getting learners' attention once they are focused on a device screen, as well as keeping them on task.

Julie's advice to teachers is:
- Don't be afraid to start using mobile devices – they can enhance your classes.
- Collaborate and share ideas with other teachers.
- Get your learners' attention back by getting them to close their devices.

For institutions, too, there are challenges. Apart from implementation challenges – such as timetabling, charging, safe storage and device management and synchronisation – there are potential issues with loss, theft or damage, and these must all be factored in to both planning and budget.

Class sets are, more than anything, prone to obsolescence, and it is a rare institution that can afford to update its technology provision with the same frequency that many individual users do.

In spite of all these challenges, class sets can represent a stepping-stone to wider deployment of mobile technologies and serve as a gentle, controlled introduction to mobile learning for both teachers and learners.

BYOD

The Bring Your Own Device/Bring Your Own Technology model (BYOD/BYOT: these terms are used interchangeably in educational technology discourse) allows each learner to bring with them whatever technology they have to hand – and to use it in class, or for learning activities on the move, outside of class.

However, in a typical BYOD scenario, the teacher might be faced with a class of learners featuring a few Apple devices, a few Android devices, perhaps a Blackberry, a Windows phone and a few low-end devices. They may also find themselves in classes where there are not enough devices to go round.

The BYOD approach is therefore ideal in scenarios where the learners are, in their majority, equipped with suitable devices such as smartphones or tablets, and where mobile and hand-held devices already play a large part in the lives of the local population.

Challenges for teachers can be found in the complexity of the 'ecosystem' (typically a mix of iOS, Android, Windows and more) and the teacher's potential unfamiliarity with certain devices, although this is often mitigated by the fact that the device owners generally know how to work their own machines, and have a vested interest in being able to demonstrate that.

There are also potential content issues, with different devices producing media across a wide variety of formats, and with apps being largely limited to the two or three main players (Apple iOS, Google Android, Microsoft Windows).

> ### Case study 3: BYOD in Argentina
>
> **Teacher:** Alicia Artusi
> **Video:** *http://youtu.be/xDyhwjcJyHY*
>
> Alicia uses a BYOD approach, and she encourages her learners to use their own mobile devices to enhance classroom activities. For example, the learners produce book summaries in digital formats using story apps, and then share these via a class blog; they download and share podcasts.
>
> She highlights the importance of classroom management when the learners are using different devices and are working on different tasks at the same time.
>
> Alicia's advice to teachers is:
> - Consider how devices can enrich your current classroom activities.
> - Guide your learners in carrying out mobile-based activities.
> - Get feedback from your learners and from trying out the tasks.

Perhaps the biggest potential challenge with the BYOD approach, however, is that of the 'digital divide', whereby some learners have suitable devices and others do not (or, indeed, do not have a device at all). Putting the learners into pairs can help with this situation, as can a hybrid model of implementation (see below). In some situations, teachers will also need to consider issues of security and cost: loss, damage or theft whilst being used in class – or outside – and of potential data charges for learners if wifi is not freely available to them.

> ### Case study 4: BYOD in Spain
>
> **Teacher:** Helen Collins
> **Video:** *http://youtu.be/EmC3sXdWxs8*
>
> Helen uses a BYOD approach, and she encourages her learners to use their mobile phones in class, particularly for audio- or video-recording communicative tasks such as speaking activities.
>
> She emphasises the importance of classroom management with these devices, and how the teacher needs to negotiate where the learners keep their phones until needed.
>
> Helen's advice to teachers is:
> - Establish clear classroom management rules for the storage and use of your learners' mobile phones.
> - Decide how to transfer recordings (files) from learner devices to yours.
> - Allow the learners to use the school's wifi network.

A BYOD implementation will often lead to compromises: compromises over app choices, over *what* can and cannot be achieved with the devices, and over *when* and *how often* they are used. To their advantage, however, they put the device firmly in the control of the learners – and the potential ability to take the learning device with them, once class has finished, is one of the major strengths of this option.

Hybrid classes

The hybrid model is based on the BYOD approach, but envisages a class set (or resource centre or library set) of devices being available to increase the number of devices where there is a shortfall in learner devices. This approach features many of the advantages and disadvantages of both models above, but perhaps the greatest advantage is that it 'levels the playing field' in class, breaking down the digital divide so that everyone has a mobile or hand-held device they can use.

With hybrid approaches where there is a surfeit of available technology, it is tempting for the teacher to put a device in the hands of each learner, but it is worth considering how much more *meaningful* practice may be when the learners are working in pairs or small groups – so individual work should only be considered when each learner has to actually produce something of their own to be assessed or marked.

The challenges of implementation

As you will have gathered by now, using mobile devices with your learners is not always plain sailing. Whilst closer to complete normalisation than any other previous educational technology, we are still some way from it, and using ubiquitous devices such as mobile phones – and, increasingly, smartphones and tablets – has not yet become a seamless part of any lesson.

This is not so much a problem with the devices themselves but, rather, with being aware of the challenges involved in *implementing* these 'new' technologies – and having some ideas about how to deal with them.

In this section, we outline some of the typical challenges that teachers and institutions face, when wanting to introduce the use of mobile devices in learning. For each of these, we also offer some solutions. We can divide the challenges into three main groups:

- Those challenges related to devices and hardware. We call these *technical challenges*.
- Those challenges related to teaching, training and content. We call these *pedagogical challenges*.
- Those challenges related to what happens in the classroom itself. We call these *classroom management challenges*.

Technical challenges

Your first technical challenge – if you are buying devices – will be connected to hardware investment and, most probably, to whether to buy Apple or Android devices.

Hardware

We could probably dedicate an entire book to looking into the pros and cons of each platform, but the basics probably centre on cost, build quality and availability of apps.
- Android tablets – particularly those with smaller screens – are often much cheaper than similar Apple devices, though their build quality may be less robust, as they can be made of plastic rather than metal.
- Android apps have been more prone to security exploits (badly-written apps gaining access to the personal information stored on your device) than Apple apps, but some people would argue that the more open app publishing policy of Google leads to more innovation and less of a 'walled garden', with Android devices being more versatile.

The best advice we can give here is to read reviews and, where possible, try any potential devices personally before making any decisions.

The hardware question is also pertinent to the BYOD and hybrid implementations, and where and how you implement mobile learning.
- If all your learners have their own devices and connectivity at home, then there is much you can do outside the classroom – in truly mobile scenarios.
- If the digital divide in your class exists to any extent whatsoever, then you will need to tailor your use of mobile devices to scenarios and activities where you can level the playing field – and this may confine your mobile projects to class, and to overlooking compulsory digital homework or activities outside class.

Software

Software can also become a challenge from the outset. If you are working in a BYOD situation, then you will be faced with the issue of deciding exactly what you can, and can't, do with the available devices.
- On a very basic level, most of the devices coming into your class should be equipped with

the basic functions – audio recorder, camera, video camera, etc – but that is really the starting point.

- Once you move on to consider the software (apps), this is where there is a great deal of potential for confusion and incompatibility across systems, although if you're only dealing with one or two operating systems (OSs) – eg Android and iOS – then you shouldn't have too many problems.
  - □ Most of the popular apps are now designed for both systems and, by choosing carefully, you should come up with an implementation plan that will cater for both devices.
  - □ Where a specific app doesn't exist for one of the systems, you will most likely find an acceptable alternative.
- The real challenge comes when other systems enter the picture. App availability for other OSs such as Blackberry, Windows, etc, is currently very limited. In these circumstances:
  - □ You can either work with shared devices in the true BYOD model – or use some of the institutional devices in the hybrid model.
  - □ You can limit the kinds of activities you do to using internal functions that all the devices share – plus web-based platforms and services.

Content

Currently, one of the biggest challenges with mobile devices is that it is very difficult to get content to learner devices and to allow your learners to share the content they produce with other learners. We have already looked at screen sharing via a variety of methods, but how do we engage with true collaborative work across devices?

The basic answer to this is judicious use of email accounts and cloud services ('cloud services' here refers to tools and services that are located online, generally on the servers of large internet companies. Cloud services can be used for file storage (eg Dropbox or Google Drive), email accounts (eg Gmail), printing (eg Google Cloud Print) – and a wide range of other purposes. For example:

Dropbox – *http://www.dropbox.com*
Google Drive – *https://drive.google.com/*
Gmail – *https://mail.google.com*
Google Cloud Print – *http://www.google.com/cloudprint/learn/*

- Start by ensuring that each of your devices has an email account associated with it, and that everyone in the group knows the accounts of everyone else – this will allow a basic communication flow from the outset.
- Next, set up a shared Dropbox (or similar) folder and invite everyone to share. With all the devices in a particular group connected to a shared folder, it becomes relatively easy to share materials you create, or materials that are created by your learners.

You can find out more about how to do this here:
http://goo.gl/OMsdgs

Another way to share files in class is to use one of the new breed of file sharing services available. We particularly like these:
Chirp – *http://chirp.io/*
(It will share files with people in a small geographical location such as a classroom.)
DeskConnect – *http://www.deskconnect.com/*
(It will allow bi-directional file sharing between computers and mobile devices.)

Alternatively, consider setting up a class community, using a tool such as Edmodo:
Edmodo – *https://www.edmodo.com/*
This will give you greater control over content, but with the added advantage of other features such as learner tracking, homework options, instant communications – and more.

The other most-sought-after capability in classroom settings is often the ability to print work that has been produced.

- On iOS devices, this can be solved easily by purchasing an AirPrint compatible printer:
  AirPrint – *http://support.apple.com/kb/HT4356*

- On Android and other devices, Google Cloud Print can often provide a workable solution: Google Cloud Print – *http://www.google.com/cloudprint/learn/*

Connectivity

Additionally, one of the technical challenges most frequently identified by teachers using mobile devices with learners is the need for robust wifi connectivity (see the case studies above). Although learners may have 2G, 3G or even 4G connections for their smartphones or tablets (if you're using a BYOD approach), these often have an associated data cost for the learner. It is preferable to have wifi in your school for your learners to use, and this will require some investment for the school.

- Although teachers of younger learners, especially, may be concerned about their using wifi, filtering systems can be put in place to *restrict* access.
- However, we would suggest that it is more useful for the learners and teachers to have *unrestricted* access to wifi, but to use this as an opportunity to learn about digital safety, and the appropriate and inappropriate use of the internet.

With learners under 18, an Appropriate Use Policy (AUP) can also be put in place (see the end of the 'Classroom management challenges' section below).

Support

Another technical challenge that some teachers face is a lack of, or inadequate, IT support. Some schools have dealt with this by having the students themselves volunteer as 'digital leaders'.

- In this case, learners make themselves available to deal with technical problems that teachers or other learners may have with devices, and they provide training (for example, in how to use certain apps) for staff and other learners.
- Involving learners as digital leaders has been effectively put in place with all ages, including at primary school level (see 'Reading' at the end of Part A).

But probably the biggest challenge is that of an institutional ban on student-owned technology.

- Some educational authorities ban the use of learners' mobile devices from the classroom *completely* – such as the state of New York (at the time of writing).
- Some schools ban the use of *any* student-owned technology, and especially of mobile phones.

Teachers working in public schools in these contexts are in a difficult position, as any use of a BYOD approach with the learners' mobile devices in the classroom would be illegal. However, these teachers can have learners using school-owned class sets of devices – and the learners' own devices outside the school – and setting some of the activities described in this book for homework.

That said, if the ban is at an institutional level, there *is* some room for manoeuvre. In these cases, the teacher can discuss with management how mobile devices might most effectively be used.

- Often, institutional bans are due to management being unable to see past some of the issues outlined here.
- Often, presenting management with an explanation of the pedagogical benefits of using mobile devices with learners, as well as also proposing a coherent and well-thought-out 'implementation plan' (see Part C), can help managers change their minds.

Pedagogical challenges

For some teachers, the main pedagogical challenge is understanding how to ensure that mobile devices are used effectively.

Effective use

'Effectively' here means understanding how and when the devices can be used to enhance language learning. As we have hopefully made clear by now, simply using mobile devices (or indeed any technology) just for its own sake is pointless. Unless the mobile-based activity is bringing some clear benefit to the learners, we are wasting our time:

- Are we increasing their motivation and engagement?
- Are we allowing them to practise and produce language in useful ways in class?
- Are we giving them the opportunity to take their language learning out of the classroom and have extra exposure to English?

Integrating many of the activities described in Part B of this book can help ensure activities are meaningful and have a clear aim. Ideally, though, the use of devices should be part of a wider implementation plan – and the support and joint lesson planning carried out with other teachers (see Step 4 of our 'Implementation plan' in Part C) will also help ensure a pedagogically sound use of mobile devices.

As a rule of thumb, every time you want to use a mobile-based activity with your learners, you need to ask yourself (and answer!):

- Why am I doing this activity with mobile devices?
- What learning benefit will this bring?

Part of using mobile devices effectively is also knowing when *not* to use them. There is still plenty of room in the mobile-based classroom for pen and paper, for board work, for drawing, for singing, and for whatever else we normally do in our language classrooms. Mobile devices are simply tools and, like any tool, they need to be used judiciously – and sometimes not used at all.

Teacher training

Teacher training and development is essential if mobile devices are to be used effectively with learners. Some schools have found that starting with a small interested group of teachers works well. This is a 'bottom-up' approach, in which teachers volunteer to become involved – rather than a 'top-down' management decision in which everybody must be involved 'like it or not' – and it can be very effective.

Teachers can be resistant to change, particularly when it involves technology:

- Making participation in a mobile-based project *voluntary* to begin with is a sensible first step, particularly in a pilot phase (see Part C).
- Having an enthusiastic team leader or 'champion' to head the mobile project with a group of teachers is also an effective way to start.

What is particularly important is that teachers are supported and developed on an on-going basis, with regular teacher get-togethers to share ideas, progress – and challenges (see above for Case study 1). If the school is using class sets of devices, such as tablets, it is a good idea for *teachers* to have one to take home as well, and to have plenty of time to become familiar with the devices and the apps – before these are introduced to the *learners*.

- Teachers – and learners, too – will have different levels of skills when it comes to using mobile devices.
- And not just teachers – but also learners – may need some help with some aspects of using devices, once these are being used regularly in the classroom.

This is not always a question of age – with students under 35 being more competent with devices, and those of an older generation being less competent.

- Younger learners may be proficient in using social networks such as Facebook, but they may have large gaps in their knowledge, such as how to use *email* appropriately.
- Teachers may find that learners even need help using *their own* devices during class at times, and this is where digital leaders can come in handy (see 'Technical challenges' above).

But generally speaking, teachers will need to have a certain number of competences and skills in place before they start using devices in the classroom, and on-going professional development must address this need.

Learner training

In some educational contexts, the learners may have negative perceptions of using mobile devices in communicative activities.

- They may have a perception that communicative activities themselves (with or without devices) are not helpful in language learning.
- They may feel, for example, that completing drill-based grammar activities is the 'best' way to learn.

To help overcome these learners' reluctance to take part in communicative activities, asking them to use their mobile devices initially with self-study apps outside of class (such as vocabulary flashcards, grammar games or pronunciation games) or to listen to podcasts can help to ease them into a more varied use of devices in the classroom later on.

- Essentially, these learners need learner training to raise their awareness of the importance of communicative activities and production in language learning.
- Starting off on familiar ground – with self-study behaviourist apps – and slowly introducing more communicative activities over time can help them start to see the benefits of alternative approaches to learning.

This is why a staged and persuasive implementation plan is so important: learner expectations and the educational context should affect what *sorts* of mobile-based activities are introduced into the classroom, *when*, and *how often*.

Assessment

Finally, the question of how – or indeed *if* – to assess the work produced by learners using mobile devices remains. In fact, this is not as tricky as it may seem. Any assessment of learner work – whether done on paper, submitted by email or created on mobile devices – needs to form part of the wider assessment procedures in the institution.

Whether assessment is on-going and continuous throughout a school term, or whether the assessment is summative (delivered at the end of the course, often in the form of an exam or test), this will affect whether learners' classwork (and possibly homework) is taken into account in assessment.

If the learners' on-going work is going to be assessed, then we can use the same criteria already in place, and add extra criteria for digital products. For example:

- The students are using class sets of mobile devices to document and describe a field trip.
- Not only the language content but also the integration of appropriate images, sourcing of quotes or images from the internet, layout, and use of hyperlinks to additional information could form an additional assessment category.

As noted above, summative assessment in most cases will be handled by more traditional means, such as final exams or tests.

Of course, as with any assessment, the learners need to know in advance exactly what is assessed and when, and against what criteria.

Of the activities described in this book, some of them lend themselves more easily to assessment than others. Typically, the longer activities described in Part C are easier to assess against clearly-established criteria, simply because learners will be generating more language – both spoken and written – in these activities.

Classroom management challenges

The first challenge concerns the physical space in your classroom.

Movement Mobile devices such as smartphones and tablets are not like laptops. They are extremely portable, and free up the learners to move around the classroom, and even the school and the surroundings.

If your classroom has fixed seating in rows, it's more difficult for students to move around using devices, and to regroup to share images or information:

- Getting the students to work in pairs, and then swapping seats to regroup, is one way around this.
- Using less structured spaces within the school itself – such as the dining room or the school library – may be another option for some teachers in some schools.

Attention With mobile devices, some teachers report that getting learners' attention away from the screen and back onto *them* can sometimes be challenging (see Case study 2).

This has been a fear with most technology advances over the years, particularly with anything that puts a barrier between the learners and each other (or the learners and the teacher): most commonly, a screen or similar.

- It is certainly the case that technology can provide easy distraction for many learners.
- This is especially true of *their own* technologies, where connections to social networks, email and similar instant communications may be a constant reminder that something potentially more interesting is happening elsewhere.

Mobile devices mitigate against this in some ways, since the screens and devices are much smaller, and the barrier greatly reduced.

Class sets of mobile devices are easier to manage for a variety of reasons – not least of which is that they can be configured not to have connections to Facebook and other social networks, and email accounts can be controlled and limited:

- If you are using tablets with covers, you can ask the students to simply close the tablets.
- For smartphones, you can ask them to put them face down on the desk in front where you can see them, or to put the phones back in their pockets (see Case study 4).

But all this must be tempered in the light of the advantages that being connected to the outside world can bring to learning.

Distraction Related to this is the challenge of keeping students on task – that is, completing the work you have set them with devices, rather than checking Facebook or texting friends.

- When using class sets of devices, this is easier – as these can contain only the apps needed for educational tasks.
- With Bring Your Own Device, keeping students on task while they are using their own devices can be more of a challenge. Setting clear time limits for tasks is one way to keep your students on task.

In the end, only *you* can decide how much distraction you will be comfortable with:

- Some teachers attempt to ban the use of tools such as Facebook for personal ends in class.
- Other teachers will let early finishers occupy their time as they see fit and – starting from that point – hold a conversation with their learners to agree on basic usage rules.

Distraction can, of course, also impact on work done *outside* class and truly *on the move*.

- Although there is nothing new in issues of motivation for autonomous learning, it must be acknowledged that these days there are more pulls on our time and more ways of getting side-tracked from the path of study.
- Although you will, as in any other implementation of self-study, encounter the same challenges with mobile learning, we do believe that good task design and a stimulating project can work in your favour.

Appropriate use Finally, an area that concerns teachers of learners under 18 is that of the inappropriate use of devices in the classroom.

- With learners under 18, class sets of devices are easier to control, in that the apps and content on the devices can be only educational in nature.
- However, many schools do use a BYOD approach, with young learners bringing their devices to class, and in these cases, schools will usually have an Acceptable Use Policy (AUP) in place.

An AUP is a document which details how devices are to be used appropriately and what constitutes inappropriate use. For more information on AUPs, see the 'Reading' section on page 28.

The big challenges

As we have seen, when we start to use mobile devices in our teaching, the challenges can be technical, pedagogical and related to classroom management.

Coping with the challenges

We can now look at two final case studies, showing how a school *teacher* and a school *manager* have successfully dealt with some of these challenges.

Given that both these educators work with *groups* of teachers, their focus is on the wider technical and pedagogical challenges, rather than the specific issues facing individual teachers in their classrooms.

These case studies suggest useful strategies for coping with the responsibility of implementing a mobile-based approach in schools – a topic which will be explored more systematically in Part C.

Case study 5: A teacher's perspective

Teacher: Jon Parnham
Video: *http://youtu.be/_DouZUpwB2Yis*

Jon's school uses class sets of tablet computers with groups of learners. The main aim of using these mobile devices is to enhance learners' language skills. When the project started, the school was keen to get the teachers involved in the project from the very outset.

Jon identifies the main **technical challenge** as ensuring reliable wifi connectivity in the school. He points out:
- Many schools focus only on the *hardware* – such as class sets of tablets.
- They pay less attention to *infrastructure* – such as connectivity, which is vital for tasks that require an internet connection.

Another technical challenge that teachers need to address when using class sets of devices is how to get the content that students produce off the tablets, before the devices go to another class.

Jon's school deals with the **pedagogical challenge** of getting teachers on board by setting up a 'special interest group' with a core group of 12–15 teachers.
- These teachers meet regularly to explore apps and tasks together, and then experiment with them in their classes.
- The teachers share ideas and resources, both in their regular face-to-face meetings and also online.

This on-going in-house teacher development has been key to the success of the project.

It is obvious from Jon's comments that there *are* challenges that need to be faced. These challenges are wide-ranging:
- Not only from the *technical* to the *pedagogical*.
- But also from the *fundamental* to what might seem the *incidental*.

In other words, it is important to take care of the major issues *and* the 'minor' details, which can also hugely affect the success of using hand-held devices. For example, how you get your learners' work off class sets of devices – and where you store that work – may seem like something that will take care of itself. It won't. You will need to plan for it.

There are challenges, then, that need to be considered.

But they are not insurmountable.

This has been a general overview of mobile and hand-held learning – and we are now more or less ready to start 'going mobile', by using the activities proposed in Part B and considering how they might work in your own context.

But first, it will be useful to outline and explain the 'model' upon which the activities are based – Ruben Puentedura's SAMR model[10].

The challenges of getting going

Going Mobile has so far examined what you need to get started using mobile and hand-held devices with your learners.
- We have looked at some of the decisions you need to take, and at some of the things you need to keep in mind.
- We can now examine the challenge of dealing with the practicalities of going mobile.

In other words, how to begin.

Focus on simplicity

Because using mobile devices in the classroom is still a relatively new phenomenon, teachers may feel unsure about where to start. Our advice is to start simple. We have divided our practical activities into five chapters:
- Each chapter builds on the last.
- Each chapter moves from the 'simplest' device functions (text), through to more 'complex' functions (image, audio and video).

In this way, we hope that teachers will be able to start at a level where they feel confident – and build from there:
- In Chapter One, the activities are less concentrated on implementing the technologies than they are on discussing them and considering their potential.
- Moving further on, we expand our repertoire of tools and utilities, and explore Pegrum's taxonomy with activities based around the three catagories discussed on page 9.

Additionally, we have identified three general foci for the activities.

Focus on connections

Your learners don't need expensive smartphones or tablet computers to carry out all of the tasks in this book. Most basic feature phones have a text function and include a camera, as well as audio- and video-recording features, but they often lack internet connectivity.

- Activities in Part B that *won't* require an internet connection can easily be carried out by learners with feature phones – an activity might simply ask the learners to collect photos with their device's camera outside of class, for example.
- Activities that *will* require a connection (whether in class or outside) need devices with an internet connection – whether that is the school wifi connection, the learners' home wifi connection, or a 2G, 3G or even 4G data connection. These activities are marked with the 'connection' icon: C

So, again, learners with basic feature phones can often still take part in 'going mobile' as well.

Focus on mobility

Most of the activities in Part B are to be done with learners in the classroom. However:

- Some of them can be done outside the classroom – for example, in your school building or grounds, at home, or even out and about at any point between school and home!
- Activities that require your learners to complete part of the task using their devices outside of the classroom are marked with the 'going mobile' icon: M

Focus on complexity

Within each chapter, as we have said, we have organised the tasks from *simpler* to more *demanding*.

But what exactly does this mean?

As a starting point, we have based the activities around Ruben Puentedura's SAMR model[10], which suggests that tasks using technology fall into two main camps:

- Tasks that use technology to *enhance* what we usually do in the classroom.
- Tasks that use technology to *transform* what we do.

The SAMR model

TRANSFORMATION

Redefinition
Technology allows for the creation of new tasks, previously inconceivable.

Modification
Technology allows for significant task redesign.

ENHANCEMENT

Augmentation
Technology acts as a direct tool substitute, with functional improvement.

Substitution
Technology acts as a tool substitution, with no functional change.

Enhancement

In the lower half of Puentedura's model, technology is used as an *enhancement* to traditional classroom activities:

■ **Substitution**

In this implementation, technology doesn't do anything that can't be done with more traditional materials and approaches. An example of this might be the note taking app, which – in a basic implementation – is a direct substitute for paper and pen. Although these apps may bring additional features, such as spell checking, etc, they are essentially an electronic version of an already existing tool.

■ **Augmentation**

Moving on from the 'notes app' example above, this implementation might use, say, a shared wiki (or Google Drive document) for process writing. Such documents allow for collaborative writing on different levels, but also allow for steps in the writing process involving peer correction, comparisons of different versions of a document, and more. As such, the technology augments – or adds to – something we might traditionally do in class.

Transformation

In the upper half of the model, technology acts as a *transformational* tool, allowing us to move beyond our traditional collection of teaching tasks and activities:

■ **Modification**

In this model, we can carry out tasks which – although conceivably based on traditional learning tasks – are significantly changed or modified by the technology. An example might be the use of multimedia presentation software to put together project work. Here, technology may change the task, allowing for a variety of approaches and steps to task accomplishment, and opening up ways for creative practice which may not have been available 'pre-technology'.

■ **Redefinition**

Here, technology takes on a more (r)evolutionary role, allowing for new ways of working and learning. An example might be the use of geolocation to provide mobile learning opportunities for students as they move through physical spaces (eg listening materials available at set points around a town, triggered as they enter each location). Such uses show technology – and our use of it – as truly transformational.

It is often easiest for us, as teachers, to start at the level of 'enhancement', where we start using technology (mobile devices, in our case) to substitute more traditional tools and to enhance the task.

Once we have gained confidence with mobile devices, and understand their special features (such as geolocation):

- We can start to introduce tasks that use these affordances.
- We can start to introduce tasks, both in the classroom and beyond the classroom walls, that would be *impossible* to carry out without mobile devices.

These are tasks that rely on the mobility of the learners, the devices, and the learning experience itself. Located towards the end of each chapter of *Going Mobile*, they require the use of a smartphone or tablet with an internet connection – and are frequently true 'going mobile' tasks.

References

1 Norbrook, H 'If They Won't Write, Get Them to Text' Retrieved 17/06/14 from *http://goo.gl/7wt50Q*

2 Kukulska-Hulme, A 'Will mobile learning change language learning?' ReCALL 21 (2) 2009 Retrieved 17/06/14 from *http://goo.gl/Pbv5n*

3 Traxler, J 'Learning in a mobile age' *International Journal of Mobile and Blended Learning* 1 (1) 2009

4 Sharples, M, Milrad, M, Arnedillo-Sánchez, I and Vavoula, G 'Mobile learning: small devices, big issues' In Balacheff, N, Ludvigsen, S, de Jong, T, Lazonder, A, Barnes, S and Montandon, L (Eds) *Technology Enhanced Learning: Principles and Products* Dordrecht: Springer 2009

5 Kukulska-Hulme, A, Sharples, M, Milrad, M, Arnedillo-Sánchez, I and Vavoula, G 'Innovation in mobile learning: a European perspective' *International Journal of Mobile and Blended Learning* 1 (1) 2009

6 Pegrum, M *Mobile Learning: Languages, Literacies and Cultures* Macmillan 2014

7 IndexMundi 'Telephones – mobile cellular per capita – Country Comparison' Retrieved 17/06/14 from *http://goo.gl/xl65L7*

8 Bax, S 'CALL – past, present and future' *System* 31 (1) 2003

9 Schrock, K 'Bloomin' Apps – Kathy Schrock's Guide to Everything' Retrieved 17/06/14 from *http://goo.gl/Zl5tVn*

10 Puentedura, R 'The SAMR Model Explained by Ruben R. Puentedura' Retrieved 17/06/14 from *http://goo.gl/Xfsx0R*

Reading

We recommend taking a look at some of the books and resources below if you would like to learn more about teaching with hand-held devices.

Brooks-Young, S *Teaching with the Tools Kids Really Use* Corwin 2010
- This book includes practical suggestions about how to integrate devices like mobile phones, MP3 players and netbooks into your classroom teaching, as well as a look at Web 2.0 tools such as social networks, virtual worlds and games.

Hockly, N 'Designer Learning: The Teacher as Designer of Mobile-based Classroom Learning' The International Research Foundation Retrieved 17/06/14 from *http://goo.gl/K8FTQi*
- This paper describes a BYOD (Bring Your Own Device) action research project carried out with international EFL learners, and describes six key parameters for effective mobile task design that emerged from the project.

Nielsen, L and Webb, W *Teaching Generation Text (grades 5–12)* Jossey-Bass 2011
- This book is aimed at secondary school teachers who would like to start using mobile phones with their students. It includes suggestions as to how to get school administrators and parents on board, as well as practical ideas for classroom activities.

Pegrum, M *Mobile Learning: Languages, Literacies and Cultures* Macmillan 2014
- This book provides an excellent overview of mobile learning and some of the cultural and educational implications. It includes several useful vignettes and case studies of mobile projects from around the world. A must read.

Quinn, C *The Mobile Academy: mLearning for Higher Education* Jossey-Bass 2011
- This book is recommended if you work with tertiary-level learners and would like some ideas about how to integrate the use of mobile devices into your teaching. Areas covered include assessment, classroom content and delivering administrative services via mobile devices.

Stacey, R 'Using Primary Aged Digital Leaders' Retrieved 17/06/14 from *http://goo.gl/DgB7wn*
- This blog post describes how one school encourages primary-aged children to be digital leaders, and gives advice to teachers wanting to create digital leaders in their classrooms.

The Webwise site 'Developing internet policy in your school' Retrieved 17/06/14 from *http://goo.gl/TJAfj5*
- This online document provides a clear and thorough guide to creating an Acceptable Use Policy (AUP) for your school, and includes useful samples and checklists.

Types of apps we have used

To carry out the activities in Parts B and C of *Going Mobile*, we recommend a variety of different types of apps.

By suggesting app *types* for the activities (rather than one single app per activity) we aim to give you some choice in what you and your learners can use, depending on the availability of the app and the devices you are using:

- Some apps we recommend are free.
- Some are currently free but may become pay-for in the future.
- Some may have a 'lite' version that is free, and a more comprehensive pay-for version.

We have tried to recommend cross-platform apps (apps that work on a range of devices) where possible. However:

- An app may only be available for one mobile operating system (eg Android or iOS mobile devices) – and we indicate this where necessary.
- An app may only be available on one platform currently, but may become cross-platform in the future – so it is always worth checking in the relevant app store.

The app types listed are arranged in alphabetical order. We briefly explain what each app type can be used for, followed by a number of specific examples:

- When you look at Part B, you will usually find app types (rather than a single app) suggested for each activity.
- When you have decided on an activity to carry out with your learners, you can come back to this list and choose a specific app from the examples given.

Whilst specific apps will come and go, all those given as examples are current at the time of writing, and we are confident that the *types* of apps we have incorporated into the activities will be available for the foreseeable future.

If an app we mention is not available when you try a particular activity:

- You can search the relevant app store for a current equivalent.
- You can use an app search engine such as Appcrawlr (*http://appcrawlr.com*).

■ ■ ■

As with all the apps recommended in this book, teachers will benefit more from a wide range used sparingly throughout a course, rather than relying only on a small set.

Animated movie apps are for creating cartoon characters with voice or text.
Examples include Tellagami (cross-platform) and Toontastic (iOS), and browser-based apps like Dvolver and Voki.

Audio recording apps allow you to record short audio files, save them online, and share the links.
Examples include Audioboo, ipadio and Cinch (all cross-platform).

Audio translation apps enable you to translate by speaking in one language and the app then translates your words into another language.
Examples include Google Translate (cross-platform), Voice Translator Free (Android) and Translate Voice Free (iOS).

Audio voiceover apps allow you to add audio to otherwise silent video files.
Examples include Dub me! (iOS) and Video Dubber (Android).

Drawing apps allow you to draw using a variety of shapes, lines and colours.
Examples include Sketch Pad 3 (cross-platform), Paper and Doodle Buddy (both iOS) and Sketchbook Mobile Express (Android).

Educational social network apps give educators a secure and more easily-controlled educational social networking space.
The most popular example is Edmodo (cross-platform).

Geocaching is an activity that gets participants to find objects (or 'caches') which have been hidden in the real world, by using GPS coordinates.
Geocaching app examples include OpenCaching, Geocaching Intro and Munzee (all cross-platform).

Geolocation apps show us the world around us overlaid with a variety of data and information (text, audio and/or video) connected to where we are.
Examples include Woices, Wikitude, Layar, Aurasma, Foursquare or Yelp (all cross-platform).

Google Voice Search (cross-platform) is a single app that allows learners to search using voice queries and commands.

Hotspot apps allow you to upload photographs and put 'hot spots' (hidden links) on them. When clicked, these hot spots can launch a variety of media: text, websites, video and audio.
A web-based example is ThingLink.

Messaging apps enable learners to send short text messages to each other, and/or images, audio and video.
Examples include Skype, WhatsApp and Snapchat (all cross-platform).

Mindmapping apps enable learners to create colourful and attractive mindmaps that can be shared.
Examples include SimpleMind, Mindmeister and Mindomo (all cross-platform) and Popplet (iOS).

Mobile games can be played on very basic mobile phones, as well as on more sophisticated devices.
Popular examples include games like Snake and Bejeweled for feature phones, and Angry Birds for smart devices. Another example is Tiny Tap, which enables learners to create their own game for smart devices with tactile screens.

Note taking apps are integrated into most phones. Some note taking apps allow you to add images, and to record audio and video from within the app.
Examples include Notability (iOS), Droid Notepad (Android) and Evernote and Microsoft Onenote (both cross-platform).

Photo annotation apps enable you to write on, and annotate or mark up, images.
Examples include Skitch (cross-platform) and PicsArt-Photo Studio (Android).

Photo collage apps allow you to create colourful collages by putting photos together in a single view.
Examples include Photo Collage, Pic Collage, Photo Grid and Pic Stitch (all cross-platform).

Photo manipulation apps are for creating amusing images of people in different contexts and situations.
Examples include Photofunia, Face in Hole (both cross-platform) and Fun Photo Box (Android).

Photo sharing apps allow you to create themed collections of images.
Examples include Flickr and Instagram (both cross-platform).

Photo story apps are for creating comic strips.
Examples include Strip Designer (iOS) and Comic Strip It! (Android).

Polling apps enable you to ask questions and collect the answers via your learners' mobile devices in real time.
Examples include Poll Everywhere, Socrative and GoSoapBox (all cross-platform).

Presentation apps are for creating engaging visual slideshows.
Examples include Keynote and SlideShark (both iOS) and myBrainshark (Android).

QR code generators enable you to create QR codes for your learners to read (see below).
Examples include Kaywa and The QR Code Generator (both web-based).

QR code readers use the camera function of your phone to read three-dimensional QR codes.
Examples include i-nigma, QR Code Reader and Scan (all cross-platform).

Screencast apps are for drawing a picture and recording a voiceover commentary at the same time.
Examples include ScreenChomp and Educreations (both iOS) and Explain Everything (cross-platform).

Social network apps are mostly used for communication, but can also be used with learners for looking at social media norms, profiles and other aspects of social media use.
Examples include Facebook (and regional equivalents, such as VKontacte or Renren) and Twitter.

Speech to text apps convert spoken words into written text by using voice recognition software. They are also known as 'dictation' apps.
Reliable examples include Dragon Dictate (iOS) and Speech to Text (Android).

Stop motion apps are for making stop motion videos with commentary.
Examples include Stop-Motion (Android) and Stop Motion Studio (iOS).

Sticky note apps allow you to create virtual noticeboards which learners can post to in the form of coloured notes.
Examples include Stormboard and Padlet (both web-based) and Linoit (cross-platform).

Synchronous (real time) communication apps enable instant communication between pairs or small groups.
Examples include Skype, WhatsApp and Facebook (all cross-platform).

Video collage apps enable you to record very short videos and share them online.
Examples include Vine, Vidstitch Free and Magisto Video Editor & Maker (all cross-platform) and Viddy (iOS).

Video image apps are for combining images/photos and audio into finished movies or audio slideshows.
Video production apps also do this (see below), and other examples include SonicPics (iOS) and Sonic Images Studio (Android).

Video production apps allow for basic video recording and editing – and more, depending on the app.
Examples include iMovie (iOS), WeVideo (Android) and Animoto Video Maker (cross-platform).

Video subtitling apps let you add subtitles to a video.
Examples include MySubtitle (iOS) and AndroMedia Video Editor (Android).

Vocabulary flashcard apps can be used for collecting, recording and recycling vocabulary.
Examples include Quizlet and the British Council's My WordBook 2 (both cross-platform).

Word cloud apps are for making visual clouds of vocabulary or other language features.
Examples include Wordle, TagCloud and CloudArt (all iOS), and Word Cloud Generator, Word Art and iLanguage Cloud (all Android).

Going Mobile is now ready to turn its attention to practical activities that you can carry out with your learners.

Here's how.

In designing these tasks, we have carefully structured them from the simpler to the more demanding, both in overall development – from Chapter One to Chapter Five – and within each chapter itself.

- Chapter One introduces the **topic** of mobile devices, and you and your learners don't need to use mobile devices at all to carry these out. This is a good place to start if you and/or your learners are new to using devices in class, or if the use of devices is not allowed in your classroom context. These activities consider mobile devices, their affordances (ie their features) and their potential application in teaching and learning. Going through them not only with learners, but also with colleagues and school managers, may help *everyone* get a bigger picture of mobile devices in education, and can work in your favour – in terms of devising policies of acceptable use, and revisiting any bans or restrictions you may have in place at work.

- Chapter Two includes activities that encourage learners to produce only **text** on their mobile devices. You will explore a variety of uses for text in a variety of genres, and try out a range of tools to either produce or consume text on mobile devices.

- Chapter Three bases the activities around the learners using their mobile devices to work with **image** as well as text. In these activities, we look at drawing, image annotation, image creation and sharing, presentations, collages and games.

- Chapter Four builds on this by introducing **audio**-based activities. We concentrate on both listening and speaking skills, with plenty of audio production. We also look at dictation activities, voice search, interviews, audio guides and more.

- Chapter Five introduces **video**-based activities, pulling together the three strands – of text, image and audio. In this final chapter, we bring together skills and apps from the previous chapters and look at how these can all be combined into more visually-rich multimedia experiences using video. The activities include video collages, voiceovers and subtitles, stop motion, video production and audio slideshows.

By the end of this part of *Going Mobile*, we hope you will have a good general knowledge of how mobile devices work, what apps are available, and how best to combine devices and apps into rich learning experiences for your classes.

Chapter One

Hands off!

You and your learners don't need to use devices at all to carry out the activities in this first chapter. That's why we've called it 'Hands off!'

If you have never used mobile devices with your learners before, or if you want a slow and structured 'way in' to using devices in your classroom, these activities are a good place to start.

As the activities are 'hands off', it doesn't matter if your learners own less sophisticated devices, or are not allowed to bring devices into the classroom or school (we have some suggestions in Part C about how to address this particular situation).

The first activity (*Features and functions*) introduces the learners to some of the basic vocabulary they need to talk about mobile devices.

This is followed by activities that encourage them to talk about mobile phones, about how they use them in their personal lives, and about various affordances of typical devices.

We also explore the downside of owning a mobile device, such as over-dependence on a mobile phone, the danger of theft, and misuse.

These are important issues that learners need to understand and be aware of, and these activities can help your learners develop their overall mobile and digital literacies.

The last two activities in this chapter (*Mobile classroom* and *Mobile rules!*) encourage the learners to explore how mobile devices can best be used in and outside the class to support their English language learning.

Holding open and frank discussions about how you intend to start integrating the use of mobile devices into your language classes is an important first step to take with your learners.

It will help them understand that there is a clear rationale and benefit to this, and will ensure that they are 'on board' – and keen to try out the activities in Chapters Two to Five.

Mobile devices are here to stay, and these activities will help pave the way towards a fuller integration of devices in your classroom practice.

This first chapter encourages you and your learners to dip your collective toes into the water!

Features and functions

A mobile vocabulary race

The learners describe a range of
mobile phone features and functions.

Run up

Prepare cards with the following words on each card, one set
of cards for each pair of learners:

texting	*hotspot*	*wifi*
ringtone	*airplane mode*	*data roaming*
contacts	*wallpaper*	*carrier*
brightness	*passcode*	*messages*
calendar	*app*	*alarm clock*
bluetooth	*keyboard*	*selfie*

Run

Show the learners a picture of a mobile phone (or show *your*
phone) and brainstorm what words they know in English
to describe the features and functions. Write them on the
board.

Put the learners into pairs, and give each pair a set of the
vocabulary cards:
- *Are any of the words on the board the same as those
  on the cards?*

Ask the learners to divide the cards into two groups:
- Those they know the meaning of.
- Those they don't know, or are not sure about.

Conduct class feedback, explaining any new words. Ensure
that everybody understands all the words. They shouldn't
write down the words for the moment (see 'Run on').

Collect all the cards.

Divide the class into two teams: A and B. Ask each team
to sit in a half circle, facing the board.

With a large group of learners, create more teams, each
with their own set of cards. A good number for each team
is about 5–8.

Give each team one set of the vocabulary cards, and ask
them to spread them out in the centre of their half circle,
face up (eg on a table, or on the floor):
- Each word should have Blutak® (or something similar)
  on the back, to allow the word to be stuck to the board.
- All the team members should be able to see clearly and
  reach the cards.

Divide the board in half (if you have two teams) and put
Team A at the top of one side and *Team B* at the top of the
other side. Allocate the members of the team a number – if
you have eight members in each team, allocate numbers 1–8
to the learners.

Tell the learners that you are going to define one of the
words, and then call out a number.

The learner in each team with that number must pick up
the correct vocabulary card, run to the board, and stick the
word there. The first learner to do this earns one point for
their team.
- Do one example with the class, to ensure they have
  understood. For example, say:
  This is the music or sound your phone makes when it rings.
- Pause for a few seconds, to allow the team to identify the
  word together.
- Call out a learner's number. For example:
  Number 5.
- Learner number 5 picks up the card with *ringtone*, runs
  to the board, and sticks the card in the space allocated
  to their team.
- Award one point to the fastest learner's team.

Continue the activity until all the cards are up on the board.

Run on

This is a lively activity that will get the learners excited,
and possibly noisy. To calm them down, you can do the
following:
- Ask them to copy the words from the board into their
  notebooks.
- Put them into pairs to write a definition or translation
  for each word.

Conduct class feedback, ensuring that everybody fully
understands all the new vocabulary.

Mobile me

Myself, my phone and I

The learners talk about their mobile phones
and what these can do.

Run up

Prepare a short presentation about your mobile phone:
- What model it is.
- What it can do.
- What you use it for in your everyday life.

Concentrate on your *personal* use.

(We will be looking at potential classroom use in
Mobile classroom on page 39.)

Run

Start by organising a 'Find someone who …' activity:

Find someone who …
- *doesn't have or use a mobile phone.*
- *has a phone that belonged to someone else in the family
  (eg their father's old phone).*
- *only takes photos with their phone.*
- *never uses the camera on their phone.*
- *has a phone with no camera.*
- *has more than five games on their phone.*
- *has had a phone confiscated in class or at home.*
- *has had a phone stolen.*
- *has more than 50 contacts in their phone.*
- *has recorded themselves on the phone.*
- *uses the phone more for messaging than speaking to friends.*
- *has used a phone in class before.*

Now talk about *your* mobile phone, using the short
presentation you prepared. Encourage the learners to ask
you questions about your mobile habits, the tools and
the apps you use.

Draw attention to the language used to describe gadgets and
their uses:
- *It's got a …*
- *I use that for …*
- *It doesn't have …*
- *I often …*
- *I don't really use the …*
- *I'm always …*

The learners describe *their* mobile phones to each other in
pairs. If some of them don't have mobile phones, they might
describe the phone features they would *like* to have.

Run on

Ask the learners to prepare a poster:
- *What is their phone like?*
- *How do they use it?*

Put the posters on the wall, and conduct a final feedback
session as a whole class.

Technology timelines

A touch of personal history

The learners create their own
'technology histories'.

Run up

Photocopy the cards on the opposite page – one complete set
for each learner in your class.

Run

Tell the learners about the technologies *you* have owned:
- The chronology.
- What they did.
- Where and when you typically use/used each technology.

Try to mention several of the technologies shown on the
cards.

Ask the learners to write down the names of the technologies
you mentioned:
- *How many can they remember?*

Elicit the names of the technologies by showing one set of
cards, and write them on the board in two columns. Before
you add them, ask:
- *Which column should each word go into?*

Older technologies	Newer digital technologies
Television	Desktop computer
Record player	MP3 player
Cassette player	Mobile phone
(or 'ghettoblaster')	Tablet computer
Walkman	Digital camera
VHS video camera	

Put the learners into small groups of three or four to tell
each other about the technologies they have owned:
- The chronology.
- Where and when they typically use/used each one.

Conduct feedback as a whole class:
- *Which older technologies do we still use today? Which
  technologies don't we use any more?*
- *Who owned some of the older technologies? What do they
  remember about them?*
- *What newer digital technologies do they use? Where and
  when?*
- *Which are the most important technologies for them
  (old and new)? Why?*

Tell the learners they are going to create their own personal
'technology histories'.

Technology timelines

Give each learner a set of the cards opposite.

Tell them they can arrange the cards (technologies) into any grouping or order, to reflect their own personal histories with these objects. For example:

- They might like to group some cards together.
- They might create a timeline showing when they acquired each one.
- They might order them from most to least important.
- They might prefer to arrange them in a circle.

Tell them that there is no one 'correct' way to organise the cards.

You can organise your own cards first to show how it can be done – for example, by sticking the cards to the board and saying:

- *I put the television and mobile phone here at the top because I use them the most.*
- *The first technologies I owned were a record player and a ghettoblaster when I was a student at university, so I put them here on the left.*
- *I use my digital camera and tablet mainly when I travel, so they are here on the other side.*
- *I've never owned a video camera, so it's not here.*

Give the learners a few minutes to work alone and arrange their cards.

In pairs, ask them to show and explain the groupings of their cards to their partner.

Conduct whole-class feedback to share the various ways of grouping the objects.

Run on

Ask the learners to write up a personal history of the technologies they currently own and/or have owned. They can organise their text anyway they like:

- Cronologically (when they acquired each).
- How each technology is used in a typical day.
- In order of importance (most to least).
- In order of size (biggest to smallest).

They could insert their technology cards inside their histories – as a way of illustrating them.

The perfect phone

My dream mobile

The learners design their perfect mobile phone.

Run up

If you did the activity *Mobile me* on page 34, ensure that you have up on the wall the posters your learners made in the 'Run on' section. These posters will serve as a good starting point for this activity.

Run

Tell the learners that they are going to design the mobile phone of their dreams.

Divide them into groups of four (A, B, C, D):

- If you have posters from the previous activity, encourage the learners to revisit them and take notes of the best features.
- If not, ask them to think about their own mobiles, and brainstorm a list of the best features for their ideal phone.

Tell them that their perfect phone can do anything they want. For example:

- It might help them with their homework.
- It might project data directly to glasses or contact lenses.
- It might play high-end video games.

They add the new features to their dream phone.

Regroup the learners into new groups of four (the As together, Bs together …) and ask them to tell each other about their new mobile phone:

- *What is it?*
- *What does it look like?*
- *What does it do?*

Each learner chooses one mobile phone to buy.

Conduct feedback as a whole class:

- *Which was the most popular phone?*
- *Why?*

Ask the learners to choose the 'top ten' features from the suggestions of the whole class and to combine them into one perfect phone.

Discuss the possible design, name, price, etc.

Run on

Ask the learners to each prepare an advertisement for the phone they designed in the initial stages of the class:

- In low-tech environments, this can be done on paper.
- In high-tech classes with the necessary skills, it can be done using PowerPoint, or as a video advertisement with commentary and music.

Share the advertisements in class, and discuss which are the most persuasive.

Addicted!

Mobile phone addiction

The learners examine and discuss their reliance on their mobile phones.

Run up

Write 'Addiction' on the board, and brainstorm with the learners what things people can be addicted to:

drugs	*power*	*alcohol*
gambling	*television*	*chocolate*
coffee	*shopping*	*computer games* …

Ask them what 'unimportant' things *they* are addicted to.

Run

Ask the learners if they think they are addicted to their mobile phones.

Dictate the following questions (or put them on the board, or hand them out):

- *Do you check your messages first thing in the morning?*
- *Do you check your phone before going to sleep?*
- *Do you check your phone regularly with no reason?*
- *Do you feel you must read a message as soon as you receive it?*
- *Do you check your phone or send text messages in social situations, eg at dinner or while talking to friends?*
- *Do you check your phone more than 30 times a day?*
- *Do you always check or use your phone during 'dead' time, such as waiting for a bus or an appointment?*
- *Do you often want to check your phone or send text messages while driving?*
- *Do you worry about losing your phone?*
- *Do you worry about being addicted to your phone?*

Ask the learners to ask and answer the questions in pairs. With lower levels, highlight the use of the auxiliary 'do' in short responses: *Yes, I do./No, I don't.*

Say that if they answered 'yes' to more than six questions, they are probably addicted to their phones!

Ask them to rate their own mobile phone addiction by placing an 'x' on a simple scale:

Not addicted ——————x———— **Very addicted**

With a small class, you can put the scale on the board and ask the learners to come up and place 'x' (or their names) on the scale.

Discuss who is the most (and least) addicted.

Run on

Challenge all the learners to give up using their phones for one day – before doing the next activity: *Cold turkey*.

Cold turkey

Dealing with addiction

The learners suggest strategies for addressing
mobile phone addiction.

Run up

If you did the activity *Addicted!* in a previous lesson, ask
the learners if they managed to give up using their phones
for one day:
- *Was it difficult? Why/why not?*
- *Is 'going cold turkey' (stopping completely) effective?*
- *Is it better to deal with mobile phone addiction more
  gradually?*

Run

Put the learners into small groups to brainstorm advice
or strategies for someone who is seriously addicted to
their mobile phone.

Say that identifying and controlling *overuse* of technology
– such as mobile phones – is part of being a healthy and
responsible citizen in today's digital world!

Offer the learners a few suggestions to get them started:
- *Take a mobile phone addiction survey to increase awareness.*
- *Every time you use your phone, ask yourself why, and if it is
  really necessary.*
- *Spend time with friends and family without constantly
  consulting your phone.*
- *Don't go cold turkey. Start to cut down slowly on your use
  of your phone.*

Other advice might include:
- *Check your phone not more than once an hour.*
- *Turn off alerts for messages and email (and smartphone
  apps).*
- *Start turning off your phone completely for short periods
  of time (eg one or two hours a day).*
- *Identify the sources of addiction (email? texting? games?
  Facebook?) and set regular times for these (eg twice a day).*
- *Delete time-wasters, like games, from your phone.*
- *Go out once or twice a week without your phone.*
- *Do activities regularly without a phone (eg reading, cooking,
  exercising).*

Regroup the learners to share their advice.
- Tell the groups to choose their *top three* pieces of advice,
  and to elect a spokesperson to share these with the class.
- Ask each group's spokesperson to present their top three.

Run on

Follow up a few days or weeks later by asking the learners
if they tried putting this advice into practice.

Safe and sound

Mobile theft

The learners make suggestions
for keeping their mobile devices safe.

Run up

Prepare to share any stories about mobile device theft
that you know:
- From your own experiences.
- Stories of mobile device theft from your friends or family.

Run

Ask the learners if anyone has had their mobile device stolen:
- *What happened?*

If necessary, elicit or explain the differences between *theft*,
robbery, *burglary*, *pick-pocketing*, *mugging* …

Share your own stories about mobile device theft:
- *How could theft have been avoided in the stories?*

Use this as an opportunity to review language for
imperatives (low levels), advice (intermediate levels) or
should have (higher levels) – by eliciting sentences onto
the board. For example:

imperatives
- *Don't leave your phone on the table in a café.*
- *Don't take out expensive devices in the street.*
- *Put a passcode on your phone.*

advice
- *You shouldn't leave your phone on the table in a café.*
- *If I were you, I wouldn't show expensive devices in the street.*
- *You could also put a passcode on your phone.*

should have
- *I shouldn't have left my phone on the table in the café.*
- *I shouldn't have taken out an expensive device in the street.*
- *I should've put a passcode on my phone.*

Put the learners into pairs. Ask each pair to come up with
a list of 8–10 sentences about how to be safe with mobile
devices, using the language you have been highlighting.

The pairs read out their sentences. As each one is read out,
the rest of the class rate them from 1–3:
- 1 = 'Not that important.'
- 3 = 'Very important.'

You all discuss their reasons.

Run on

Put the learners into pairs again:
- They write down, from memory, all the sentences rated 3.
- Conduct class feedback:
  How many pieces of advice can each pair remember?

Ask the learners which advice they will try to put into practice
themselves, to keep their own mobile devices safe in future.

Don't do it!

Mobile conflicts

The learners roleplay scenarios
of inappropriate mobile phone use.

Run up

Prepare five roleplay scenarios based on situations where
mobile phones are not used appropriately, or use those
suggested opposite:

- For each pair of learners in the class, create complete sets
  of all the scenarios.
- For each scenario, put the roles for A and B on separate
  pieces of paper.

Run

Ask the learners to think of annoying ways that people use
mobile phones. For example:

- *How do they feel when someone has a loud phone
  conversation on the bus?*
- *Do they get annoyed if someone they are talking to
  is constantly checking their phone?*

Put the learners into pairs to brainstorm a short list of
inappropriate ways to use mobile phones.

Conduct class feedback, and add the ideas to the board.

Tell the learners they are going to roleplay five situations
with inappropriate uses of mobile phones.

Put them into pairs, and allocate roles A and B to each pair.

- Give each pair a set of scenarios 1–5, with the roles on
  separate pieces of paper.
- They keep each set of scenarios *face down* for the moment.

When everybody is ready, they pick up their role (A or B) for
scenario 1 and read the roles.

Give them a minute or two to prepare their own role, and
then ask them to do the roleplay. Remind them to be polite!

You listen and note any polite phrases you hear them use.

Conduct feedback, putting the polite phrases you heard
on the board, and adding the phrases below if necessary:

- *You really shouldn't …*
- *I'd prefer you not to …*
- *You're not allowed to …*
- *I don't like the way you are …*
- *Can you please stop (+ …ing)?*

If you write up this language *after* the learners have roleplayed
scenario 1, they use the language they already know. If they are
fairly low-level, you could provide it *before* the roleplay.

Ask the learners to proceed with scenarios 2–5. Monitor the
pairs, noting any typical language errors.

Conduct class feedback, highlighting any language errors.

---

1 The cheating student

A: You are a student, and you are using your mobile
phone during an exam. Phones are banned during
exams. Prepare some good excuses!

B: You are a teacher. You see A using their phone during
an exam. Confiscate A's phone.

2 The show-off

A: You come from a very rich family, and have been given
the latest, most expensive tablet for your birthday.
Try to make B jealous of your new device.

B: You have a very basic mobile phone. A is going to
show off about their new device. Show them that you
are not interested or impressed.

3 The movie-goer

A: You are in the cinema, and your phone rings. It's your
mother on the line. She has been ill recently, so you
answer the call. She's a bit deaf, so you need to speak
loudly.

B: You are in the cinema. The person sitting next to you
is having a loud mobile phone conversation. Tell them
you are not happy about this.

4 The arguing couple

A: You are in a restaurant having a meal with a friend.
Your friend keeps looking at their phone and sending
text messages while you are talking. Tell them you are
not happy about this.

B: You are in a restaurant having a meal with a friend.
You had a serious argument with your partner before
coming out, and are trying to fix the situation by text
message. If you leave it until after the meal, you know
your partner will be even angrier.

5 The bully

A: You have received an unpleasant text message from
someone in your school. B is your best friend. Tell
them about it, and ask for advice.

B: A is your best friend. A has received an unpleasant
text message from someone in your school. Give
them some advice about what to do.

Run on

Ask the pairs to create a roleplay scenario card for one more
inappropriate use of mobile phones. They can refer to their
brainstormed ideas from earlier. They write their roleplay on
a piece of paper, using scenarios 1–5 as a model.

- Redistribute the new scenarios.
- Ask the learners to carry out their new roleplay.

Conduct class feedback, asking the learners to share the new
scenarios, and discuss the outcomes of their roleplays.

Mobile classroom

What can we do?

The learners talk about potential uses
of mobile phones in class.

Run up

Prepare a short anecdote about your educational uses of
your mobile phone – this might include:

- Reading blogs or websites.
- Downloading videos or podcasts.
- Interacting with other teachers on Facebook or Twitter,
  or similar.
- Watching TED videos (*http://www.ted.com/*).

Run

Tell the learners about your learning uses for your mobile,
divide them into small groups and ask them to do the same.

They may not come up with too many uses:

- If mobile phones are banned in class.
- If they don't consider their mobile activity to be
  connected with learning.

If they do have trouble at this stage, you may want to go
directly to the table opposite.

Conduct feedback as a whole class, then tell the learners
you are actually considering using mobile phones for their
language learning.

Ask them to think about what their mobile phones can do.
If you have done *Mobile me* on page 34, they could look back
at the features of their phones while they brainstorm in pairs.

Either on the board or projected on screen, elicit common
features of mobile phones to the *first* column of a table like
the one opposite.

The learners think how they might use these features in class,
and complete the *second* column of the table. You may need
to give them a few examples like the ones opposite.

Conduct whole-class feedback:

- *Which activities and features would they most like to use in
  their English classes?*

Run on

Ask each learner to choose one aspect from the completed
table and plan a week's activity around it. For example:

- Record two vocabulary items each day, using the camera.
- Practise speaking each day, using the audio recorder.

After one week, conduct feedback. Ask the learners to share
what they did:

- *How useful did they think it was?*
- *What did they learn?*

Ask them if they enjoyed it and would like to do more of it.

Digital camera	
Audio recorder	
Video camera	
Note-taker	
Maps	
Voice memos	
Apps	

Digital camera	*Take pictures of new vocabulary items.*
Audio recorder	*Pronunciation practice.*
Video camera	*Video speaking activities for language analysis.*
Note-taker	*Note down new words and phrases.*
Maps	*Practise directions.*
Voice memos	*Record short pieces for fluency practice.*
Apps	*Dictionaries, translation apps, presentations, ebooks …*

Mobile rules!

A contract for classroom use

The learners prepare a 'fair use' policy
for mobile phones in class.

Run up

This is a good activity to finish with, before moving on to
further activities in Part B.

- If you have already done the previous activity on page 39
  – *Mobile classroom* – you will have some idea of what the
  learners will be *able* to do with their mobiles, and what
  they would *like* to do with them.
- If not, prepare a chart like the one in *Mobile classroom*,
  with typical mobile phone features and possible classroom
  uses.

Run

Remind the learners of their proposed educational uses for
their mobile phones in class, or display the chart you have
prepared:

- *Do they have anything to add to the chart?*

Tell them that, although there are plenty of good reasons to
use mobile phones in class, there are also several potential
drawbacks.

Put the learners into pairs to brainstorm these drawbacks.

If you think they will have trouble getting started, give them
a few ideas:

- They might use Facebook during the lesson.
- They might rely too much on the devices
  (eg for translation).
- Some mobile phones may run out of battery.
- Some may be lost or stolen.

Conduct feedback as a whole class.

Tell the learners that they are going to agree a contract
for 'acceptable use' of mobile phones in class.

Discuss each potential drawback:

- *Is it something that should, or should not, be allowed?*
  (eg using Facebook during a lesson)
- *Is it something that needs a solution?*
  (eg phones running out of battery)

Explore possible solutions, or ask the learners to justify
their decisions.

As the class discussion progresses, make notes on the board
under two headings:

Things not allowed during class time	Things to remember
Talk to friends on Facebook.	*Come to school with a full battery.*

When the learners have agreed the final content of the table,
draw it up as a contract for future mobile use in class.

You and the learners all sign the contract.

Run on

Display the contract on a wall.

- The learners make notes of the rules in the note-taker
  on their device, for future reference.
- They illustrate some of the rules and pieces of advice.
  (eg *No Facebook!*)

Post the illustrations around the classroom.

Chapter Two

Text

Hands on!

Chapter Two focuses on a device feature that is very familiar to most and is not technically challenging: the text or note taking function of mobile devices.

However, the text-based activities in this chapter are not just about writing. They integrate the range of skills, as well as grammar and vocabulary.

The activities that require writing focus on getting the learners to create short texts that reflect the way we use our mobile devices in the real world when we write (or more accurately, type): to create short SMS messages or notes, and to type short phrases and individual words.

And if your learners are going to use mobile phones in class (as opposed to tablets), short text-based activities are far more suited to the limited screen size and very small keyboards of these devices.

The first two dictation activities are a simple and effective way to start using mobile devices in the classroom. They don't require any special apps or an internet connection, and they can be carried out with very basic mobile phones.

These are followed by activities that do require an internet connection, and get the learners to produce words, short sentences or short texts.

For example: they add virtual sticky notes to an online noticeboard, produce word clouds, and contribute to a real-time online poll on their heroes.

The learners then explore, analyse and use a range of public and private social networks, to support their language learning – Facebook (or the local equivalent) in *Networks*, Edmodo in *Mr. Ed* and Twitter in *Twitter celebrities*.

Contributing to social networks is one of the most common uses of mobile devices today, and your learners may well already do this regularly in their own language.

The next three activities get the learners to work with the language of texting and with emoticons – something, again, that they also very likely do, and enjoy, in their first language.

The chapter ends with two activities which can form the basis for ongoing revision: creating and using vocabulary flashcards, and using mindmaps for note taking and review.

After the 'hands off' activities of Chapter One, this chapter will help you to start working with devices in the classroom – via text-based activities that can be fitted easily into your syllabus.

C Connectivity

This icon at the top of an activity indicates that at some point the learners will need to be connected to the internet.

M Mobility

This icon indicates that at some point in the activity the learners will be asked to 'go mobile'.

Know your letters

Phone features

The learners review spelling, vocabulary and language through three short dictation activities.

Run up

Basic mobile phones can be used for these activities.
Check whether everybody in the class owns a mobile phone. If not:

- Ensure that there are enough phones for the learners to share.
- Ask them to take turns using the phones to complete the activities.

Prepare the three short dictation activities opposite:

- *Letters*
- *Mini-texts*
- *Questions*

Run

Tell the learners you are going to give them three short dictation activities – using their mobile phones.

Follow the procedures outlined opposite.

Run on

Extend the activities, by doing a slightly different kind of dictation:

- Dictate another five questions to the learners, but instead of typing in the *questions* themselves, as in Activity 3, they type the *answers* into their mobile phones.
- Put the learners into pairs.
- Ask them to recreate the original *questions* by looking at their *answers*, and to type the questions into their phones.

Conduct feedback, by eliciting the questions onto the board – so that the learners can check their spelling/language.

Activity 1: Letters

Prepare by writing down four or five sentences giving instructions to do something simple. For example:
Put both of your hands on top of your head.
Stand on one leg.
Put your left arm on your right shoulder.

To do the activity, tell the learners they are going to take down a short dictation of individual letters in their mobile phone SMS text messaging function, or in a note taking app if their phone has this:

- Tell them that the letters join together to form a sentence that asks them to do something. As soon as they understand the sentence, they should carry out the action.
- Look at the first sentence you created earlier and read it out – one letter at a time:
  P-U-T-B-O-T-H-O-F-Y-O-U-R-H-A-N-D-S …

Don't pause much between letters, and don't repeat letters.

As soon as the learners understand the sentence, they do the action:

- You finish dictating all the letters.
- The learners compare their texts:
  Can they break the stream of letters up into individual words?

You then repeat the process with the other sentences you prepared.

Put the learners into pairs:

- Ask the pairs to write down one or two new sentences – with instructions to do something simple.
- Ask them to change partners, and to dictate their instructions to their new partner, letter by letter.

As soon as they understand the sentence formed by the dictated letters, the new partner carries out the action.

Know your letters

Activity 2: Mini-texts

Prepare by choosing a short text (about 25 words) relevant to the interests and level of your learners – this could be an extract from the coursebook, for example.

To do the activity, tell the learners they are going to take down a short dictation in their mobile phone SMS text messaging function (or preferably in a note taking app, if their phone has this):

- You read the short text you chose earlier at normal speed, while the learners just listen.
- They then open the text messaging function (or note taking app) on their phones.
- You repeat the text slowly as a dictation – so that the learners have time to type it in.
- They then check in pairs that they have the same text, by comparing their dictations.

Elicit the text from the class and write it up on the board, highlighting any issues that arise with spelling or language.

Ask the learners to look through the coursebook units you have already studied:

- They choose a short extract of about 25–30 words from any of the reading texts.
- In pairs, they dictate their chosen mini-texts to each other.
- They then check their dictations with the original text in the book.

Conduct feedback with the whole class:

- *What did the learners find challenging about this activity?*
- *What did they find easy?*

Activity 3: Questions

Prepare by noting down five questions to ask the learners on a topic they have studied recently. For example:

- *What's your favourite movie?*
- *How often do you go to the cinema?*
- *Do you prefer dubbed or subtitled movies?*

To do the activity, tell the learners they are going to take down a short dictation of five questions in their mobile phone SMS text messaging function (or preferably in a note taking app, if their phone has this):

- The learners open the text messaging function (or note taking app) on their phones.
- You read out each question slowly, so that they have time to type them all in.

Elicit the questions from the class:

- You write them on the board.
- You highlight any issues that arise with spelling/ language.

Ask the learners to work in pairs, and to ask and answer each question as a short speaking activity.

Conduct feedback with the class, so that they can briefly compare and discuss their answers.

Know your numbers

More phone features

The learners review numbers and dates through
three short dictation activities.

Run up

Basic mobile phones can be used for this activity.

Check if all the learners in the class own a mobile phone.
If not, ensure that there are enough phones for them to
share. They take turns using the phones.

Prepare three short dictation activities – see below.

Run

Tell the learners you are going to give them three short
dictation activities – using their mobile phones:
- *Numbers*
- *Maths*
- *Dates*

Follow the procedures outlined below.

Run on

Get the learners to share their birthdays:
- Put them into groups of five or six, and ask them to share
  the dates of their birthdays with their group.
- They all type the name and birthday of their group
  members into their mobile phone calendars. If your class
  is small (maximum 15 learners), do this as a whole class.
- Conduct feedback:
  *How many people in the class have birthdays in each month
  of the year?*

Make a note yourself of the learners' birthdays, so you can
congratulate them on the day!

Activity 1: Numbers

In a previous lesson and with their permission, the
learners give you their phone numbers each on a
separate piece of paper.

On the day of the activity, tell the learners they are
going to add their classmates' phone numbers in their
mobile phones:

- Ask them to open their contacts lists/address books
  on their phones.
- From the list of phone numbers you collected
  previously, read out each learner's name, followed by
  the phone number – once only – while the learners
  type it in.
- Ask them to check in pairs that they have noted down
  all the numbers correctly, by taking turns to read each
  other's number out loud from their phones.

This provides them with additional oral practice.

Activity 2: Maths

Prepare by noting down 10 maths sums (and answers)
which practise large numbers:

What's 7917 plus 708?   *What's 1048 minus 416?*
What's 15 times 312?   *What's 30,900 divided by 3? …*

To do the activity, tell the learners they are going to solve
10 maths problems, using their mobile phone calculators.

Check that they know the English words for these
four simple mathematical operations, by putting them
on the board:

+ (plus)   x (times / multiplied by)
– (minus)   ÷ (divided by)

Ask them to open the calculators on their phones:

- From the list of maths sums you compiled, read out
  each one in full, while the learners type it into their
  calculators.
- Ask them to check in pairs that they have the same
  answers, by reading them out (quietly!) in pairs.

Elicit the answers from the class, highlighting any
language issues that arise with the numbers.

Activity 3: Dates

Prepare by noting down 8–10 important national dates:
- *Independence Day*
- *Public holidays*
- *Religious feast days …*

To do the activity, tell the learners they are going to add
important national dates to their mobile phone calendars.

Ask them to open their calendars on their phones:

- From the list of national dates you compiled, read out
  each date in full – followed by the day/event – while
  the learners type it in:
  The 17th of July is Independence Day.
  The 25th of December is Christmas Day.
- They check in pairs that they have noted down the
  dates and events correctly, by taking turns to read
  each date out to each other from their phones.

This can open out into a general discussion about
'famous dates'.

Sticky boards

Five facts

The learners add five personal 'facts' about themselves to an online noticeboard.

Run up

Set up an online noticeboard, using a sticky note app.

Add five short personal 'facts' about yourself to an online sticky note.

Include facts your learners are unlikely to know about you. For example:

- *I started learning to play the piano aged 10.*
- *I have visited three continents.*
- *I love cooking Indian food.*

If you would like to review a specific language area (*can*, the present perfect …) include only these structures in your facts.

If you plan to do the 'Run on' activity, create a second separate online noticeboard.

Run

Show the learners the online noticeboard, and ask them to read the sticky note with your five facts:

- *Did they know any of these things about you already?*
- *Which of your facts do they think are the most unusual or interesting?*

Encourage them to ask further questions about each of the facts (review question forms with the class, if necessary).

Tell the learners:

- They are going to each add five short personal facts about *themselves* to the noticeboard.
- They must *not* add their names.

Give them the noticeboard URL, and enough time to write and save their five facts in online sticky notes.

Put them into pairs. They read each sticky note:

- *Which one was written by their partner?*

The partner confirms or corrects their choice.

Still in their pairs, tell the learners to guess:

- *Which of the other classmates do they think wrote each of the sticky notes?*

Conduct class feedback.

For each sticky note, tell the learners to ask that particular learner further questions about each fact.

Run on

Use a second online noticeboard to give the learners more reading and writing practice, and for them to get to know each other better:

- Assign each learner one of their classmate's names, and ask them to carefully re-read their 'five facts' sticky note.
- Give them a five-minute time limit to look for one image, and one video, that they think this classmate will enjoy, based on their five facts.
- Give them the URL of the new noticeboard you created before the lesson.

Tell the learners to add to a new sticky note on the new online noticeboard and include:

- Their assigned classmate's name.
- Their chosen image.
- A link to their chosen video.

Ask them to find the sticky note addressed to them:

- They look at the image.
- They watch the video.

Conduct class feedback on the choices of images and videos:

- *What images and videos did their classmate choose for them?*
- *Were they relevant to their five facts, and if so, how?*
- *Whose images and/or videos were the most unusual or interesting?*

There are a lot of different types of activities to do with online noticeboards:

- You can get feedback from the learners on a class, an idea, or an issue.
- You can use the noticeboard to brainstorm ideas as a class, or as a place for the learners to write definitions of words or ideas.

You can also encourage the learners to write very short texts on a class-related topic, and to include an image or a link to a relevant video in a single sticky note.

In the clouds

Words in the sky

The learners experiment with word clouds.

Run up

Ensure that you and your learners have a word cloud app on your devices.

The best way to introduce the topic of word clouds is to use one in context as part of a lesson – we suggest creating one as an introduction to a reading or listening text.

As an example, below is a word cloud (made with the Wordle app) from a text about Barcelona in Spain.

Run

Show the learners the word cloud you created, and ask them some questions about it.

In our example, we ask the following:

- *We're going to read a text about the place where Nicky lives. Where do you think it is?*
- *That's right, it's Barcelona. Where is Barcelona?*
- *Right – it's in Spain. Is it a big place?*
- *Yes, it's a city. What else can you guess about the city?*

Before moving on, make sure that the learners know the following:

- This kind of image is called a word cloud.
- The size of the word in the word cloud reflects the frequency of the word in the original text. This is an important feature, as it can help them focus on key words and concepts in a text.

Give the learners some time to look through your word cloud and make some more guesses.

- Then continue with your reading or listening lesson.
- At the end, use the word cloud again as a review aid.

Show the learners how you made your word cloud, and ask them to create their own clouds from a text.

Make sure they are conscious of any extra options:

- *Fonts*
- *Colours*
- *Layouts …*

Run on

Assign a recent topic to each learner in the class:

- Ask them to make a word cloud of related vocabulary items for homework.
- You can then use the word clouds in class for topic or language revision.

Word clouds can be shared as images, but can also be printed and hung in class as visual reminders of content covered during a course.

There are many ways of exploiting word clouds in – and out of – the language classroom.

As noted, they are excellent for pre-reading and post-reading exercises, but they are also good for:

- *Vocabulary revision*
- *Thematic lessons*
- *Research topics*
- *Grammar points (eg word order)*

And much more!

Image created in Wordle, from a text here: *http://goo.gl/umKYkl*

Heroes

Polling

The learners share opinions on their heroes
in a mobile poll.

Run up

Find images of 8–10 famous people and characters. For example:

- *Superman*
- *Robin Hood*
- *Shakira*
- *Ghandi*
- *Serena Williams*
- *Albert Einstein …*

Choose people your learners will recognise, and include a range of professions.

Set up two polls in a polling app:

- Poll 1 is a multiple-choice poll, and includes the question *Who is the hero?* and the names of each famous person.
- Poll 2 is an open-response poll, and includes the question *What is a hero?*

Run

Ask the learners in pairs to complete the sentence:
A hero is someone who … .

Ask the pairs to read out their sentences for the class.

Show the learners the images of your chosen famous people. Elicit who each one is, and why they are famous:

- *Are these people heroes?*
- *Why/Why not?*

Ask the learners to briefly discuss this in their pairs.

Tell them they are going to vote:

- *Which famous person do they think is most 'a hero'?*

Project Poll 1 onto the board:

- The learners use one mobile device in their pair to choose one famous person.
- The poll results will update in real time on the board.

When they have finished voting, look at the poll results on the board as a class:

- *Which person gets the most votes?*
- *Which person gets the least votes?*
- *Why?*

Now tell the pairs to discuss the characteristics of a hero. Give them two minutes to think of as many characteristics as possible.

Elicit or give them a couple of examples to start them off:

- *A hero thinks of other people first.*
- *A hero contributes something important to mankind.*

Tell the learners they will contribute their characteristics to a poll.

Project Poll 2 onto the board:

- The learners type their hero characteristics one by one.
- They send them to the poll.
- The poll page updates in real time on the board.

Once the learners have finished sending their characteristics, slowly scroll back through the contributions in the poll, so that they have time to read them all:

- *Which characteristics are the most common?*
- *Which characteristics are the most unusual or original?*
- *Which characteristics do they agree with most/least?*

Highlight some of the language errors in the contributions. Because polling contributions are anonymous, individual learners will not feel singled out by this.

Ask them to correct the errors in pairs, then check as a class.

Run on

You can extend this activity to provide writing practice:

- Ask the learners in pairs to choose someone they think is a hero. Give them a five-minute time limit to find information about their chosen person online.
- Ask them to prepare a 100-word text about why this person is a hero. The texts can be written on paper, or in a texting or note taking app on the mobile devices.

Share the learners' completed texts by putting the papers on a noticeboard or on the walls of the classroom, or by asking them to upload their electronic texts to a class blog.

If you did the previous activity – *In the clouds* – you could ask your learners to create word clouds from their texts, and share those.

Networks

Interactive language practice

The learners post to a public social network page for language learners.

Run up

Before class, visit several public Facebook pages aimed at English language learners, and choose one appropriate for your learners. For example:
The British Council LearnEnglish page for adults:
http://learnenglish.britishcouncil.org/en/
LearnEnglish Teens and LearnEnglish Kids for young learners:
http://learnenglishteens.britishcouncil.org/
http://learnenglishkids.britishcouncil.org/en/

If Facebook is not commonly used in your context, choose the most popular social networking site (eg VKontakte in Eastern Europe, Renren in China, etc) and try to find equivalent public language sites in English.

Check how many of the learners have Facebook (or VK or Renren, etc) accounts:
- Statistically, it is probable that at least half the class will have an account on a social network.
- If fewer than half the learners have accounts, put them into small groups with one social network user per group.

Run

Introduce the topic of networks. Ask the learners to discuss the following questions in small groups:
- *What real-life clubs or groups do they belong to (sports club, choir, reading group, dance class, etc)?*
- *What do they enjoy about belonging to these groups?*
- *What is the most popular social network in their country?*
- *Do they have an account? Why/Why not?*
- *Do they personally know everyone in their online social network(s)?*
- *What do they enjoy about belonging to an online network?*
- *What language(s) do they use in their online social network?*
- *Is belonging to an online network as good as belonging to a real-life network? Why/Why not?*

Conduct class feedback on these questions:
- Some learners may not like or want to use social networks.
- This provides a good opportunity for class discussion.

Take your cue from the learners, and if the majority of the class are not interested in online social networks, then perhaps stop the activity at this point!

Point out that there are social network pages for English language learners:
- These can provide extra language practice.
- They also provide the chance to interact with teachers and learners around the world.

Show the learners the public language page you chose before class (eg the British Council LearnEnglish page(s) on Facebook or local equivalent):
- They log on with their mobile devices.
- They find and 'like' the page you have just shown them.

Ask them to explore the postings on the page, and to discuss the following questions:
- *How many people like this page?*
- *When was the page created?*
  (Tip: Look at the page timeline.)
- *What is the latest posting about?*
- *How many people have commented on this latest posting so far?*
- *Which posting do they think is the most interesting, of those they can see on the main page?*
- *Which posting is the least interesting for them? Why?*
- *Whose comments are the most interesting? Why?*
- *Whose comments are less interesting or valuable? Why?*

Tell the learners to note down:
- One thing they like about the page.
- One thing they think the page can do, to help them improve their English.

Conduct feedback.

Ask the learners to choose one posting from the past week, and to reply to it. Help them with language and vocabulary.

Run on

Set the learners an out-of-class 'language challenge':
- Ask them to add a short daily comment to the social network page, for one week.
- One week later, ask them:
  Were their postings to the site commented on?
  If so, by whom?
  Where did they post from – home, school, on the move …?

Ask them if they enjoyed the language challenge:
- *Will they continue reading the posts from this social network page?*
- *Will they continue posting their own responses now and again?*

Mr. Ed

Educational network profiles

The learners create an account on an educational social network.

Run up

Set up a teacher account at Edmodo (*www.edmodo.com*) and create a group for the learners to join.

Get the learners to install the Edmodo app on their devices.

Ensure that you have filled in your own profile, as a model for the learners to copy.

Run

Take some key words from your own Edmodo profile, and put them on the board.

Put the learners into small groups:
- They try to construct your profile from the key words.
- They can do this on paper, or in a note taking app on their devices.

Bring the class back together and reconstruct your profile on the board together, using their suggestions.

When you have a finished model, project or show your real Edmodo profile and compare:
- *What did they get right?*
- *What is missing?*

As a class, brainstorm what makes a good profile in an online social space:
- *What information is vital?*
- *What is desirable?*
- *What should not be included?*
- *Is there a difference between a purely social network (like Facebook) and an educational one?*

Ask the learners to write their own profile for Edmodo:
- They do it in a note taking app.
- They will then be able to copy and paste it into Edmodo.

Conduct peer correction in pairs, and deal with any questions they may have.

Give the learners the 'join code' for your Edmodo group, and ask them to sign up and join the group:
- Ask them to copy the profiles they have written and paste them into their profiles in Edmodo.
- Ask them to read the others' profiles and comment on anything they find of interest.

Run on

Ask the learners to access Edmodo before the next lesson, encouraging them to explore the options in the community:
- Ask each learner to post a question about Edmodo to the group, and to answer any that they can help with.
- Answer the remaining questions yourself.

If you set up Edmodo accounts and a group early in a course, you can use it throughout the term or year:
- For extra practice.
- For sharing URLs.
- For messaging.

And more – it becomes, in effect, a private social network space for the class.

Twitter celebrities

Tweeting the famous

The learners follow international and
local celebrities on Twitter.

Run up

Sign up for Twitter, if you don't already have an account.

In your Twitter account, choose a few international (and local) celebrities to follow:

- We don't recommend here specific individuals for you to follow – fame can be fleeting, and today's popular singer or sports personality may be an unknown tomorrow!
- We do suggest you ask the learners to choose celebrities *they* are interested in following.

Instead of individuals, you could propose these categories – to help the learners search for celebrities on Twitter:

- *Singers*
- *Bands*
- *Actors*
- *Sports personalities*
- *TV personalities*
- *Popular scientists*
- *Authors*
- *Politicians*
- *Business leaders*

Run

Introduce the topic by showing a few pictures of celebrities, including both international and local celebrities.

- *Do your learners recognise them?*
- *Do they follow any celebrities on Twitter?*

Put this (invented) example tweet on the board:

Excited abtgiving concert tonite in Seattle! @ladygaga doing
a guest appearance w me 2 support #WWF charity

Ask the learners:

- *What is this?* (a 'tweet' – a Twitter message)
- *Is this tweet from an actor, singer or politician?* (a singer)
- *What is this singer doing tonight?* (a concert in Seattle)
- *Who is also going to sing?* (Lady Gaga – a famous singer)
- *What is the concert for?* (to raise money for charity)

Elicit and highlight the following characteristics of the tweet:

- It is short (tweets have a maximum of 140 characters).
- There are abbreviations and 'textspeak' spellings – *w* for *with*, *abt* for *about*, *tonite* for *tonight* – although tweets are also sometimes written in standard English.
- There are numbers for words (*2* for *to*).
- @ladygaga is Lady Gaga's Twitter name (the @ symbol shows this).
- # is a hashtag – a way to label and search for tweets by key word or topic.
  (#WWF is the hashtag used to label information about the World Wildlife Fund.)

Ask the learners to download the Twitter app onto their devices, and to sign up for a Twitter account if they don't already have one.

If the learners are new to Twitter, and/or they are reluctant to open accounts:

- Open a single class account.
- Give everyone the username and password.

Ask the learners to search for four or five international and/or local celebrities to follow.

Put them into pairs, and ask them to choose one of the celebrities they are now following:

- They look at the last few tweets from this celebrity.
- They try to understand them.

You help them, as necessary.

Tell the learners to choose a recent tweet from their celebrity, and to 'retweet' it.

Before the lesson ends, ask the learners to share their own Twitter names, and to follow you and each other.

Run on

Extend this activity, by encouraging the learners to use their Twitter accounts outside of class to practise their English.

- Ask them to check their Twitter accounts daily over the next week, and to retweet any celebrity tweets that they enjoy or find interesting.
- Tell them they can also create and send their own tweets, using standard English.
- Make sure that *you* do the same during the week!

In the next lesson, the learners share their experience of using Twitter:

- *How often did they check their Twitter account on their mobile device during the week?*
- *What did they learn about the celebrities they follow?*
- *How many celebrity tweets did they retweet?*
- *What did they retweet?*
- *Did they send any original tweets, and if so, what?*
- *If not, would they like to try tweeting?*
- *Where did they use Twitter during the week (at home, work, school, on the move …)?*
- *Do they think Twitter can help them improve their English? Why/Why not?*

Using their Twitter accounts outside of class can give the learners extra exposure to authentic English, and following celebrities they are interested in can be very motivating.

Short and sweet

Text messages

The learners compare short 'text message' English
with 'standard' English.

Run up

Prepare pieces of paper with sentences in standard English.
Make sure you have one sentence for each pair of learners.
For example:

- *Thanks for a great party last night. The food was absolutely
  delicious!*
- *Can you pick up the newspaper for me on your way home?*
- *Let's go to the movies later. Shall we meet on the corner of
  5th and Main Street?*
- *The meeting tonight is at 7.00 pm. Please bring that
  document with you.*
- *I can't talk to you right now because I'm in class.*
- *My teacher said my essay wasn't good enough. I failed
  the exam.*

Run

Ask the learners about their use of texting. For example:

- *How many text messages do they send and receive every day?*
- *Do they use mobile phone text messages, or do they use free
  texting apps like Whats app?*
- *Does their phone have 'text prediction', to help with spelling?*
- *Do they use text abbreviations and emoticons when they
  text? For example?*
- *Do they use capital letters and punctuation? For example?*
- *What are the differences between 'text' language and
  'standard' written language?*

Remind them that text language is very flexible, and is often
shorter than standard written forms – we often leave out
unimportant words.

Put the following message on the board:

- *Don't forget that we have a test next week, and that you will
  need to study hard!*

Ask the learners to suggest a short text message version of
this, and add their suggestions to the board:

- *What would the shortest possible version be?*

They should suggest something like:

- *Test next week – study hard!*

Ask the following:

- *What words can we leave out in text messages?*
  (Pronouns, conjunctions, articles, even short phrases)
- *What words do we need to include in text messages?*
  (Content words that carry the meaning – usually nouns,
  adjectives and adverbs. These can be abbreviated – for
  example, *week* can be *wk* – but if we use text prediction,
  we often include the whole word as it's quicker to select
  that than to type out an abbreviation.)

Put the learners into pairs.

Give each pair one of the messages in standard English
that you prepared earlier.

- The pairs create a shortened text message version on
  separate piece of paper.
- They exchange their shortened messages with another
  pair.

Tell them to read the shortened text message they have
received from the other pair:

- They write it out in standard English.
- They remember to add missing words to create
  complete sentences.

Each pair compares these standard English versions with
the originals, then shares the differences between the longer
versions and the shorter versions with the class.

As a class, examine any grammatical/language differences:

- *Are they correct, or not?*
- *Do they change the original meaning?*

Run on

Repeat this procedure, this time with the learners working
on their own:

- Each learner prepares a standard English sentence on a
  piece of paper, or in a text message, with content they
  would typically find in a text message.
- They could use one of their own real messages, translated
  into English.
- They exchange their standard sentences with another
  learner, and create 'short text' versions. These versions can
  be created on paper or on their phones.
- They then exchange their short text versions with a
  different learner. If the text versions were created on their
  phones, they can text them to their new partner.
- They change the text versions back into standard English.

Finally, the class compare the standard English translations
with the originals, and discuss any differences.

Txtng

Deconstructing text English

The learners decode an essay written in text English.

Run up

Look at the 'text language' essay in the 'Run' section below. For more on the original story behind it, see here: *http://goo.gl/ikl8Ow*

Ensure you have handouts (or PowerPoint slides or similar) with the text English and plain English versions of the essay, to show the class.

Run

Introduce the story, by telling the learners that in the UK it is customary, on the first day back to school after the summer holidays, for learners to write a short essay about what they did during the summer.

Tell them that you are going to show them one such essay, from a Scottish student.

Put the first line of the essay up on the board or screen:
My smmr hols wr CWOT. B4, we usd 2go2 NY 2C my bro, his GF &thr 3 :-0 kds

Ask for suggestions as to what it means:
My summer holidays were a complete waste of time. Before, we used to go to New York to see my brother, his girlfriend and their three screaming kids.

Now highlight some of the features of text English:

smmr	Sometimes we leave out letters from a word if it can still be understood.
CWOT	Three-letter and four-letter acronyms are common in text English.
2go2	We can use numbers when they sound the same as a word.
2C	We can use letters for the same thing.

Put the learners into small groups and hand out (or project) the whole text for them to look at. See opposite.

- In each group, ensure there is a secretary.
- The secretary will take notes in a note taking app on their mobile device.

Bring the class back together again and conduct whole-class feedback:

- Ensure that everyone has the correct story (see the Key).
- Elicit any other features of text English that they spotted.

Now give each of the learners a sample sentence to rewrite in text English. Here are some suggestions:

- *See you at eight tonight for a drink?*
- *Would you like to go to the cinema at the weekend?*
- *Where did you go for your last summer holiday?*
- *Did you do the homework yesterday?*

Text

My smmr hols wr CWOT. B4, we usd 2go2 NY 2C my bro, his GF &thr 3 :-0 kdsBt my Ps wr so {:-/ BC o 9/11 tht they dcdd 2 stay in SCO &spnd 2wks up N. Up N, WUCIWUG -- 0. I wsvvvbrd in MON. 0 bt baas & ^^^^^. AAR8, my Ps wr :-) -- they sd ICBW, &tht they wr ha-p 4 the pc&qt…IDTS!! I wntd 2 go hm ASAP, 2C my M8s again. 2day, I cam bk 2 skool. I feel v O:-) BC I hvdn all my hmwrk. Now its BAU …

Key

My summer holidays were a complete waste of time. Before, we used to go to New York to see my brother, his girlfriend and their three screaming kids. But my parents were so upset because of 09/11[1] that they decided to stay in Scotland and spend two weeks up north. Up north, what you see is what you get – nothing. I was very, very, very bored in the middle of nowhere. Nothing but sheep[2] and mountains. At any rate, my parents were happy – they said it could be worse, and that they were happy for the peace and quiet. Idiots! I wanted to go home as soon as possible to see my mates again. Today I came back to school. I feel very saintly[3] because I have done all my homework. Now it's business as usual …

[1] September 11, when the attack on the World Trade Center happened.
[2] 'Baa' is the sound a sheep makes in English.
[3] The emoticon has a halo.

Ask them to write the text version of their sentence, using a note taking or drawing app:

- The learners can take turns to project (or write) their new sentences on the board or screen.
- Their classmates guess what the original version was.

Run on

Play is a common feature of language use. Ask your learners to keep an eye out in town for examples of incorrect – or interesting – uses of English (eg a hairdresser called 'A Cut Above') and to make a note (or take a photograph).

Use the examples in class for correction, or to examine the 'language play' elements.

Face off

Emoticonversations

The learners create conversations
from emoticon prompts.

Run up

Get a series of emoticons printed out and put them in an envelope or bag for each group of learners.

A collection like this one is a good starting point: *http://goo.gl/FPZINj*

Ensure your learners all have a messaging app installed.

Run

Show the sequence opposite (on the board or screen) – with the emoticons in place, but *not* including the *text*.

Use the emoticons to elicit a conversation between the man and the woman. The answers opposite are simply a possible scenario.

Divide the class into groups, and give them a set of emoticons each:

- They work together.
- They draw one emoticon at a time.
- They decide on its meaning.
- They put together a conversation in a note taking or messaging app.

After a given time, bring the class back together and give each group a chance to display their conversation.

This is also a good time to work on error correction.

Run on

This activity can be done as an extension activity for homework, in pairs, using any standard messaging app:

- Give each pair their emoticons.
- Alternatively, ask them to build up a conversation around a recent theme or language point, using emoticons of their choice.

This is an open-ended activity that can be used regularly for recycling functions, social English – and a lot more. It gives the learners a chance to be creative with the language.

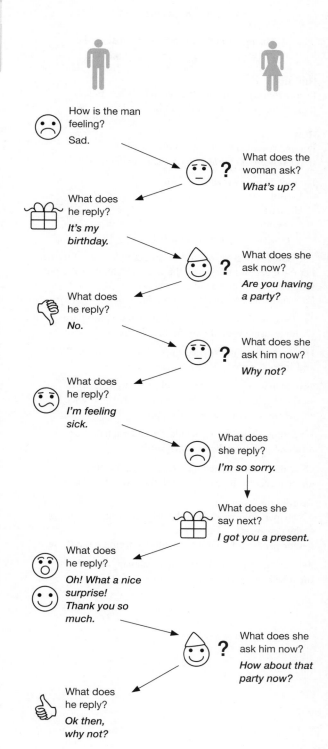

Very flash

Vocabulary revision aids

The learners create vocabulary resources.

Run up

Ensure that you and your learners have a vocabulary flashcard app installed on the mobile devices.

Run

Put one vocabulary area up on the board:
- *What words do the learners associate with the area?*

For each item, try to elicit the following:
- A word
- The part of speech
- An example sentence
- A translation – where appropriate

Use Google image search to find an example image for each item – or get examples from the Creative Commons resource ELTPics, which can be found on Flickr: *http://www.flickr.com/photos/eltpics/sets/*

Tell the class they are going to create a new vocabulary resource, using your chosen flashcard app.

Divide the class into small groups (Group A, Group B …) and appoint a group leader for each one.

Give each group a recent vocabulary set to work with:
- They brainstorm the words they know from the set.
- They add them as a set to the app.

If possible, they should include a photograph and/or an audio pronunciation of each word.

Once they have finished their collections, ask the group leader to share the set with the rest of the class.

Reorder the groups:
- One person from each of the original groups in each of the new ones.
- One person from Group A with one from Group B, etc.

Ask each member to share their vocabulary set with the others, using the app.

Run on

The learners can make their own sets of words outside of class:
- You can encourage them to collect their own sets, using photographs they take themselves.
- You can give each learner a topic, to research and to prepare a set.

The vocabulary sets can be shared and recycled in class at a later date:
- The learners can review the vocabulary sets, using the built-in flashcard capability.
- You can also review them through a variety of class-based or homework activities.

Here is an activity to get you started:

Collages

Divide the class into teams and give them a lexical area each:

- They select the 10 most useful words from that lexical area.

- They turn them into a collage, using word processing, presentation or photo collage apps.

These collages can be shared and imported into photo annotation apps and labelled by each group of learners, as a vocabulary recycling task.

All in the mind

Mindmapping

The learners create mindmaps on classroom topics.

Run up

Ensure that the learners have a mindmapping app installed on their devices.

For more versatility, a note taking app that allows for the addition of images, etc, may suit your needs more.

Run

Start by modelling the construction of a mindmap yourself, eliciting items from the learners and adding them.

You may want to concentrate on areas such as the following:
- A recent vocabulary area.
- A topic for writing or speaking.
- A theme you have recently covered in class.

Below is an example mindmap, based on a vocabulary set connected to transport.

When you have collectively added one or two items to the mindmap, divide the class into small groups and give each group a branch of the mindmap to complete:
- Give the groups time to brainstorm concepts to add.
- Then ask them to come to the board to elaborate on the mindmap.

At this stage, you may want to add other options, such as illustrations or photographs, to each item – these may help some learners internalise the vocabulary better.

Now ask the learners to choose a recent vocabulary area you have covered in class, and to create their own mindmaps with their mobile devices, using a mindmapping app.

Put them into small groups:
- They share the mindmaps.
- They then upload them to a class blog.

Conduct a class discussion about the usefulness of this kind of visual aid:
- *Do the learners like them?*
- *What do they think they could use them for?*

Here are some suggestions:
- Grammatical elements.
- Research for a talk or paper.
- Vocabulary revision.
- Book or film reviews.
- Lesson content.

Run on

Mindmaps can serve as excellent revision tools:
- Assign one lesson to each learner, and ask them to make a mindmap during the lesson, using their mindmapping app or another note taking application.
- Get them to share their mindmaps on an Edmodo group – see the activity *Mr. Ed* on page 49.

These mindmaps can then be used for further revision activities.

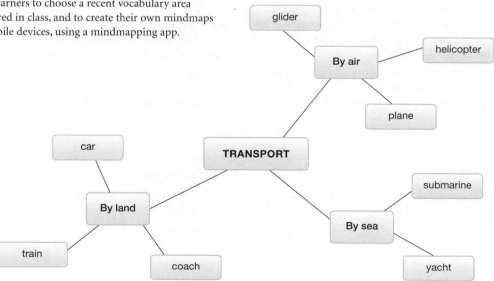

Chapter Three

Hands on!

Image

In Chapter Three, we continue with our staged approach to introducing the use of mobile devices in the classroom.

The activities in this chapter focus on the learners using or creating images alone, or images plus text.

In this way, we build on Chapter Two (text) and we get the learners to start using the camera function of mobile devices – something which even very basic mobile phones have.

And we continue to keep things simple in technical terms (taking photos is not technically challenging) so that we can keep the focus in our lessons firmly on the most important thing – language.

Many of the activities in this chapter require the learners to use their mobile devices to take photos – in the classroom or school, at home or around town – and to use these photos to take part in a range of language activities.

We start with a game activity in which the learners share photos they already have on their devices and get to know more about each other.

The next six activities have learners taking photos to practise language (eg modal verbs of probability in *Close-ups*) and vocabulary (lexical sets in *Find it* and *Word bank*). Photos by the learners are then used as a springboard for communicative speaking, listening and writing practice.

Images are combined with photo notation and/or comic strip apps in the following two activities: *Photo fit* and *All around me*. And the following activity – *Picture this* – uses a drawing app, so that the learners create their own image from scratch in response to a story.

The final two activities in this chapter are slightly different, as they don't use standard photos or images. *The name of the game* explores the imagery in simple mobile games that can be played on basic mobile phones, and *In the 'hood* provides an introduction to using geolocation apps (explored further in Part C) based on local landmarks.

Although this final activity is not based directly on photos or images, famous landmarks usually generate thousands of photos!

C Connectivity

This icon at the top of an activity indicates that at some point the learners will need to be connected to the internet.

M Mobility

This icon indicates that at some point in the activity the learners will be asked to 'go mobile'.

Got it!

My photos and me

The learners share personal photos, as a team building or 'getting to know you' activity.

Run up

This activity works with basic mobile phones.

Check that at least half the learners in the class own mobile phones.

Ensure that for each group of four or five learners, there are at least two mobile phone owners.

Run

Put the learners into groups of four or five:
- They take out their mobile phones.
- They get ready to share some of their personal pictures.

Call out a category from the example categories opposite (eg *a pet*).
- The first student in a group to find a photo of that category cries *Got it!*
- You award that group one point.

One member from each of the other groups must now ask one question about the photo being shared.

First, review question forms and add prompts to the board if necessary. For example:
- *Where is this?*
- *Who is this?*
- *What's his/her name?*
- *How old is he/she?*
- *Why is this photo important to you?*
- *What do you like most about this photo?*

Award one extra point for each question *asked* correctly and *answered* correctly.

Tell the learners to take notes during the photo sharing – they will need this information again later.

Continue the activity by choosing another category:
- The first student in a group to find a photo of that category cries *Got it!*
- You award that group one point.

Continue with the 'question and answer' procedure as above.

Repeat the procedure several more times with different categories.

- *A pet*
- *A grandparent, parent or child*
- *A good friend*
- *A group of friends*
- *A celebration*
- *A holiday photo*
- *A photo taken in nature*
- *A photo taken in your home*
- *A photo of you doing a sport*
- *A 'selfie' (a photo of you taken by yourself)*
- *A photo of a view*

Run on

Extend the activity in class by getting the learners to share more photos:
- Ask them to share five of their own photos with their group, and to ask and answer questions about the photos within their groups.
- Ask them to each choose one of their group members, and to write a short biography about that person, based on what they have learned through the photos.

If you have a class blog, the learners can upload, share and comment on each other's texts for homework.

Acknowledgement: This activity is based on one which is described here:
http://goo.gl/giRR76

Close-ups

Guess the photo

The learners take close-up photos of objects, and guess what they are.

Run up

Use a mobile device to take 6–8 close-up photos of everyday objects. For example:

- A watch
- A door hinge
- A light bulb
- A pen
- Stationery
- Foodstuff
- Clothing
- Furniture

The photo should be close enough to make it difficult (but not impossible) to recognise the object. See opposite for some close-ups taken by Nicky of a plug, a door handle, an office chair and a stapler.

Put your photos where you can easily share them with the class. For example:

- Print them out.
- Put them in a presentation slide.
- Put them on a blog page.
- Upload them to a photo sharing site such as Flickr.

Run

Show the learners one of your close-up photos:
- What do they think it could be?

Encourage them to use modal verbs of probability to guess and, if necessary, put the language on the board. For example:
- It might be a clock.
- It could be a machine.
- It may be a digital watch.
- It can't be a bracelet.

Show the learners in pairs the rest of your close-up photos:
- What do they think they could be?

After a few minutes, conduct open-class feedback:
- Encourage the learners to use modal verbs of probability to guess what each photo is.
- Where necessary, tell them what each object is.

Ask the pairs to use one mobile device to take four or five close-up photos of objects in and around the classroom or the school.

Give them a clear time limit (five minutes should be enough).

The learners exchange their mobile device with another pair:
- What do they think the photos could be?

Conduct open-class feedback:
- What was in the close-up photos?
- What objects were photographed by several pairs?

Run on

Ask the learners to take close-up photos of 5–8 classmates' eyes only. (Tell them to turn off the flash on their devices when they do this!)

Put them into pairs, and ask them to show each other the 'eyes' photos:
- Can they guess who each classmate is from the eyes alone?

Encourage them to use modals of probability:
- It must be Susanna because she's got blue eyes.
- It can't be Erica – her eyes are brown.

In some contexts, it may not be appropriate for the learners to photograph each other's faces or eyes, so you could suggest they photograph each other's hands – or shoes – instead.

Acknowledgement: Thank you Julie Cartwright for the idea for the 'Run' part of this activity – see Case study 2 in Part A on page 16.

Find it!

Photo treasure hunt

The learners take photos of items
and create treasure hunts.

Run up

Prepare a worksheet with a description of five things for the learners to find and photograph in the school building and grounds. See opposite for some ideas,

This treasure hunt reviews colours and shapes – make sure you first check that there are objects like this in the school for the learners to photograph!

You could choose to teach or review other lexical sets in your treasure hunt:

- *Clothes*
- *Physical appearance*
- *Furniture ...*

Run

Put the learners into pairs and ask them to use one mobile device with a camera per pair. Tell them:

- They are going to take part in a treasure hunt.
- They are going to find and photograph certain things.
- They are going to have a time limit of 10 minutes.

Hand out your worksheet, and give the learners time to read it. Point out:

- They can photograph anything they want.
- They need to agree, as a pair, that their chosen object matches the description on the worksheet.

Start the activity:

- The learners leave the classroom.
- They find and photograph objects in other parts of the school or grounds.

Give them a 10-minute limit to take their photos and return to class.

Back in class, regroup them into groups of four, with each learner from a different pair:

- They share each of their photos with their group.
- They explain what worksheet item each photo fits.
  For example:
  This leaf is small and green.

Conduct feedback:

- *What things did the class photograph for each of the items in the worksheet?*

> ### Ideas for things to find and photograph
>
> 1 *Something small and green.*
>
> 2 *Something white in the shape of a rectangle.*
>
> 3 *Something in the shape of a triangle.*
>
> 4 *Something with three colours.*
>
> 5 *Something big and brown.*

> ### Ideas for a treasure hunt
>
> - *Objects with geometric shapes.*
> - *Objects with certain colours, patterns and/or textures.*
> - *Classroom stationery.*
> - *Furniture.*
> - *Clothing.*
> - *Parts of the body.*
> - *Objects with certain smells and tastes*
>   *(if the learners have access to a school cafeteria).*

Run on

Put the learners into new pairs. Each pair writes a treasure hunt worksheet for the others, with five items to find and photograph in the classroom or school.

- They should include items with a common theme.
- You put some ideas on the board to help them. See above.

Conduct class feedback and discuss what items were found and photographed.

Word bank

Photo vocabulary box

The learners create their own picture dictionary
on Flickr.

Run up

For this activity, you will need to create a Flickr account
for use in class:

http://www.flickr.com

When you have created the account, set up a Flickr group:
http://www.flickr.com/groups/

Add the learners to it, and create a couple of sets of photos
of common vocabulary areas:

- *Animals*
- *Food*
- *Drink* …

Upload a couple of your own photos to each set. Make sure
that you label each photograph with the vocabulary item
and a description. You can see some good examples here:
http://www.flickr.com/photos/eltpics/sets/

You can download and install the free Flickr app – and ask
your learners to do the same, if you plan on working within
an app rather than a mobile browser.

Run

Choose one of the sets from ELTPics that reviews a lexical set
you have been working with recently:
http://www.flickr.com/photos/eltpics/sets/

Brainstorm around the set as a whole class:

- Write the words up on the board.
- Ensure that all the learners are familiar with the words
  that are generated.

Put the learners into small groups, to visit the lexical set
on Flickr:

- They have to find photo examples of the words they
  reviewed on the board.
- They have to find two new words in the set on Flickr.
- They then share them with the rest of the class.

Tell them that they are going to start compiling their own
picture dictionary of new vocabulary and examples of
'grammar in use' as they occur:

- They will need to take a photograph of any important
  new words or phrases they encounter, and upload them
  to the Flickr group.
- They will need to ensure that they add the new word,
  a sample of its use and (in monolingual classes)
  a translation.

Once you have the resource in place, make sure you refer
to it and use it regularly – so that the learners see the value
of contributing to it:

- Having this kind of personalised resource can make the
  learning of new language much more meaningful.
- Having an online vocabulary book gives plenty of creative
  opportunities for recycling key vocabulary with creative
  writing or speaking activities, quizzes, etc.

Run on

Your learners can contribute to the Flickr sets from home,
or when they are out and about in town:

- This can help them review and learn vocabulary outside
  of class time.

You can use the Flickr resource across classes at the same
or similar levels.

It can also be an excellent resource for learners, once they are
no longer with you.

My other me

Photo manipulation

The learners create an alternative life story
with manipulated photos.

Run up

Before class download and install a photo manipulation app
on your mobile device, and ask the learners to do the same.

- Use the app to create five or six photos of yourself.
- These should represent a fictional life before you became
  a teacher, so be creative and make it fun. See opposite for
  some manipulated photos of Gavin!
- Add the photos to a single slide, using a presentation app.

Run

Put the learners into pairs, and ask them to look at your
fictional life-story photos:

- Tell them that these photos represent a time in your life
  before you became a teacher.
- Tell them to work in pairs and put the pictures in order,
  deciding exactly what happened to you.

If necessary, review language for making deductions, giving
opinions, etc:

- *Perhaps he was …*
- *I think he …*
- *Maybe he …*

Divide the pairs again, so that each learner is sitting with
someone new.

They tell each other their stories and compare their versions.

Conduct feedback as a whole group:

- First share some of the stories.
- Then tell them your fictional story, based on your
  manipulated photos.

Show the learners:

- How you made the photos, using your chosen photo
  manipulation app.
- How you saved them.
- How you put them together into one slide.

Ask each learner to do the same:

- They create a fictional story for a part of their life.
- They add them to a single slide.

Put them into pairs:

- If they are using tablets, ask them to exchange devices.
- If they are using mobile phones, the small screen size
  makes it difficult to see, so they should first upload
  their slides to a shared space online.

Each learner asks their partner questions, to find out
the fictional story behind the photos.

Conduct whole-class feedback and listen to the stories.

Run on

The photos created by the learners can provide a
springboard to a range of possible activities involving
writing, reading, speaking and listening.

The learners write up their own fictional story:

- They can write it up as a newspaper or magazine article:
  'The lives and times of …'.
- They illustrate it with the original photos.

The learners conduct an audio or video interview with their
partner, as if it were a radio or TV interview, using the audio
or video recorders in their mobile devices.

Share the media files as a whole class, and vote on:
the best/most interesting/most creative.

Encourage the learners to share their work in a class blog,
or in a similar shared online space such as Edmodo (see the
activity *Mr. Ed* on page 49).

Picture puzzle

Photo stories

The learners jointly construct a story
from a set of random photographs.

Run up

Before this activity, ask each learner to come to class with a
photograph they've taken somewhere outside class:

- *A street scene*
- *A person*
- *An object …*

You bring a photo yourself, of a general scene-setting
photograph of a location.

Although this activity can be done with simple tools such as
a presentation or annotation app (see the *Photo fit* activity
on page 64) you may want to try something more creative
and get the learners to install a 'comic strip' app on their
devices.

Run

Start off by describing and showing *your* photo.

As you describe it, the learners make some brief notes
about it on paper or in a note taking app on their devices.

Nominate learners in the class:

- Each one describes and shows *their* photo.
- The others make notes.

Once all the photos have been described, put the learners
into small groups to order the pictures into a story, using
their notes.

When each group has finished, feed back as a whole class
and vote:

- *Which story is worth developing into the 'final' version?*

Now lay out the devices around the class, putting them
in order according to the chosen story.

As a whole class, negotiate what is happening in each photo:

- *What are people thinking, doing or saying?*
- *How does it fit within the story?*

Assign a photo to each learner, and show them how to add
notes to the photo.

If you are using a photo story or comic strip app, you
will be able to add speech and thought bubbles, as well as
traditional scene-setting descriptions.

As a whole class, review the story, with each learner adding
their part as you work through it:

- Encourage them to put some feeling into the story.
- Encourage them to make the telling as dynamic
  as possible.

Run on

Ask the learners to write up the story for homework:

- They can use the images and make it into a comic strip
  themselves.
- They can write up the original story *they* thought of (if it
  was not the one chosen for development by the class).

Most comic strip apps will allow you to export the finished
stories as PDF files, which can then be shared in a class blog
or similar and used in class for further language work,
error analysis, etc.

Time will tell

Photo collages

The learners create a collage of photos of objects for a time capsule.

Run up

Make a collage of photos for your own time capsule before the lesson, using a photo collage app. See Gavin's collage opposite.

Alternatively, create the collage as a single slide, using a presentation app.

Explain the concept of a time capsule to the learners, and ask them to come to class with some photos of objects for their own capsule:

- Show them your example, to give them some ideas.
- Help them to download a photo collage app.

They take the photos when they are out and about before the following lesson.

Run

Project your photo collage, or share it with the learners as a printed handout:

- Explain that these are the objects *you* would put in a time capsule – to be buried for future generations.
- Explain to the learners that they have to find out why you've chosen these items for your time capsule – by asking questions about the photos.

Put the learners into small groups, and ask them to look at the photos and brainstorm some questions for you.

- When they are ready, conduct a whole-class 'question and answer' session, giving them more information about the objects you chose – and why you chose them.

Show the learners how you created your collage, and ask them to do the same with the photos they have brought to class.

- When they are ready, pair them and ask them to interview each other about their photographs.

As the learners conduct their interviews:

- They can take notes in a note taking app.
- Alternatively, they can record the interview, using an audio or video recorder on their device.

Regroup the pairs, and ask them to tell their new partner what they have learnt about the time capsule of their interviewee.

Run on

Ask the learners to choose the most important photo from their collage and prepare a 'show and tell' presentation for a future lesson. This can be done using a presentation app.

In future lessons, give each learner some time to share their presentation and to answer questions from the class.

If you are short of time in class to share presentations, you may want to investigate the Pecha Kucha format:

- It limits presentations to a maximum of 20 slides.
- Each slide is timed to run for 20 seconds.

You can find out more about Pecha Kucha from the official website: *http://www.pechakucha.org/*

See also *Show and tell* on page 90 for a Pecha Kucha activity.

Photo fit

Photo descriptions

The learners mark up photos
using a photo annotation app.

Run up

Before this activity, ask the learners to download and install
a photo annotation app.

Find a photo of a large group of people and put it in a
shared online space so the learners can easily download it.

Ensure that each learner has the photo on their device
and imported to the annotation app.

Run

Put the learners into pairs and ask them to open up the
photo in the annotation app:
- Ask one learner of each pair to choose a person
  from the photo.
- Ask the other learner in each pair to find out
  who they have chosen, by asking questions:
  Is it a girl?
  Does she have long hair?
  Is she wearing glasses? …

As the learners ask questions and get answers, they cross out
the people in the photo who don't fit the description – and
annotate *why* on the photo itself (eg *She has long hair …*).

Once they have found the chosen person. They 'undo' the
notes they have made on the photo:
- They exchange roles.
- They repeat the activity.

Run on

When the learners have done one activity using a photo
annotation tool, they very quickly get the hang of it – and
it can be used regularly for image and vocabulary-related
activities, based on their own photographs.

This allows for a wide variety of image-related activities
in one single app. Opposite are a few, to get you started.

As you experiment with these ideas, you will no doubt begin
to think of new ones of your own.

Picture dictionary

Remembering a lot of new vocabulary items can be
difficult and stressful, particularly at beginner and
lower levels.

- Visual cues can be a powerful reminder.
- With a photo annotation app, the learners can label
  different items within a photo or sketch.
- They can be used later as a visual reminder.

Exam preparation

Many examination speaking tasks involve describing
or sharing pictures in pairwork activities.

- A basic photo annotation app will allow one learner
  to describe a photograph, and their partner to draw it.
- This is also useful for working with particular elements
  of the language – such as colours, clothes, etc – and
  even grammatical elements, such as prepositions.

Spot the difference

An extension of our main activity here – and another
useful activity that can be done for exam practice – is a
'spot the difference' photo activity.

- This involves two sketches or photographs which are
  very similar, but not the same.
- Loading one image on one device and the other on
  a second one will allow the learners to work in pairs
  to spot the differences, through detailed description
  of their photos.
- As with the exam preparation activity outlined above,
  this is particularly useful for image-specific vocabulary
  (*in the foreground*, *at the top on the left* …) but also
  for more general vocabulary related to descriptions.

All around me

English spoken

The learners examine examples of English
from their own environment.

Run up

Take some photographs of English in use in the town where
you live and teach. Try to get a balance of correct and
incorrect English usage.

This might include:

- *Shop names*
- *Shop signs*
- *Street signs*
- *Restaurant menus*
- *Tourist information …*

If it is going to be difficult to find examples around you,
you can try searching for them online. Look for:

- 'Bad English signs'
- 'English signs'
- 'English menus' …

You can find examples of bad English usage here:
http://www.engrish.com/
Or you can use the example signs opposite.

Run

Tell the learners you went for a walk yesterday (or did an
internet search) and collected examples of English – some
good ones and some bad ones:

- Give them printouts of the photos, in pairs.
- If you are using a class set of devices, you might like to
  synch a copy of the photos to each device and conduct
  the rest of this activity on the devices themselves.

Ask the learners to divide the photos into two sets:

- Those that are correct.
- Those that have errors.

When they have finished, put each pair with another pair
to make groups of four, and ask them to compare their
two sets and – if necessary – produce one final set of correct
and incorrect examples to work with:

- The learners go through the incorrect set and make the
  necessary corrections.
- If you're using synchronised photos on a class set of
  devices, you could use a photo annotation app and get
  the learners to correct the images in that. (See the
  previous activity *Photo fit* on page 64 for more on
  photo annotation.)

Conduct feedback as a whole class, identifying the incorrect
examples and working through the corrections.

Run on

Ask the learners to collect examples of English over the
following week, using the cameras in their phones or
tablet computers, or by searching on the internet.

You can then work with these in a subsequent lesson,
along similar lines to the above.

Collect the images and save them to a Flickr account:
http://www.flickr.com

They will make a useful personalised visual dictionary and
phrase book for the learners (see the activity *Word bank* on
page 60 for more).

Picture this

Drawing the story

The learners listen to a story and draw a picture
to summarise or interpret it.

Run up

Find a short story with a clear narrative, to read aloud to
your learners. Choose a story with an appropriate linguistic
level and which will take around 3–5 minutes to read.

You could choose a text from your coursebook, from a
graded reader, or from a short story website.

This activity uses a modern poem version of the fairy tale
'Little Red Riding Hood' by Roald Dahl, available online:
http://goo.gl/Xe8wS

Read through your chosen text and identify 10–15 key
vocabulary items to pre-teach.

Run

Write *Little Red Riding Hood* on the board, and elicit the
four main characters (Little Red Riding Hood, Grandma,
the Wolf, the Woodsman).

Elicit the basic original storyline from the learners.

While the learners are telling you the original story, ensure
that the following key vocabulary items are elicited and
write the words on the board:
- *The woods*
- *Basket*
- *Red cloak*
- *Grandma's cottage*
- *Grandma's clothes*
- *Furry*
- *Ears*
- *Eyes*
- *Teeth*
- *Axe*

Also write the famous lines from the fairy tale:
- Little Red Riding Hood to the Wolf:
  What big ears/eyes/teeth you have!
- The Wolf to Little Red Riding Hood:
  All the better to hear/see/eat you with!

This is also a good time to review narrative tenses, especially
the past simple and past continuous.

Put the learners into pairs, and tell them to open a drawing
app on their devices.

They take turns:
- One draws one of the vocabulary items from the board.
- Their partner provides the word.

Tell the learners you are going to read them a modern poem
version of Little Red Riding Hood, which has a very *different*
ending.

Read the poem version of Little Red Riding Hood by
Roald Dahl, available online at: *http://goo.gl/Xe8wS*.

You may want to mime some of the actions, so that the
learners fully understand what happens in the story.

Give the learners a few minutes to draw a scene from the
poem version of the story, using a drawing app on their
devices.

Put them into groups of four, to compare their drawings
and to describe their scene:
- *Did everyone draw similar or different scenes from the story?*
- *Which scene did most of them draw?*

Run on

Ask the learners to upload their drawings to a class blog,
with a short text describing what happened in the scene.
Encourage them to use narrative tenses (past simple and
past continuous) in their texts.

Roald Dahl wrote several other modern poem versions of
fairy tales (*Cinderella*, *The Three Little Pigs*, *Goldilocks and
the Three Bears*), all of which can be found online.

If the learners enjoyed this activity and the poem, you could
repeat the process with another of his tales.

With young learners, you could use traditional fairy tales
or short children's stories.

The name of the game

Mobile games

The learners teach each other how to play their favourite mobile phone games.

Run up

Download a popular game to your mobile phone, such as Snake or Bejeweled, and learn how to play it yourself.

- *Snake* is suitable for basic mobile phones.
- *Bejeweled* is best played on a feature phone or smartphone.

Opposite are some suggestions for describing Snake. (*Snake* is one of the most popular mobile phone games of all time. It first appeared on phones in 1998.)

Run

Show the learners your mobile phone and tell them you are going to describe your favourite game, without telling them the name of the game.

Describe the game, using modal verbs of obligation:

- *Do they know or recognise the game from your description?*
- *Have they ever played it?*

Elicit some of the sentences you used to describe the game, and write them on the board. Alternatively, tell the learners about the game again, this time as a dictation.

Highlight:

- The key vocabulary used to describe mobile phone games:
  screen
  arrow keys
  score points
  high score
  the game ends
- The modal verbs of obligation you used:
  have to
  can
  must
  mustn't

Put the learners into groups of three or four:

- Ask them to take out their mobile phones, and to choose one game to describe.
- Give them time to prepare their descriptions in writing, if necessary.

Tell them to describe their favourite game to their group members – without mentioning the name of the game:

- *Do the group members recognise the game?*
- *Have they played it before?*
- *If not, would they like to play it?*

Playing Snake

- *You have to move a snake around your phone's screen.*
- *You can control the snake with the arrow keys.*
- *There is a piece of food on the screen.*
- *The snake gets longer when it eats the food.*
- *You can score points when the snake eats a piece of food.*
- *You mustn't let the snake touch its own tail.*
- *You must try to get a high score by making the snake as long as you can.*
- *The game ends if the snake touches its tail.*

When all the groups have shared their game descriptions, conduct class feedback:

- *What games did they describe?*
- *How many games were new to them?*
- *Which are the most popular games?*
- *Are there any new games they would like to try?*
- *When do they normally play mobile phone games?*

Run on

Ask the learners to exchange phones with a member in their group, and to play the game described by that member.

Give them a time limit for this (about five minutes).

Alternatively, they can write a description of this game and how to play it, using modal verbs of obligation:

- They must also add one thing they like about the game, and one thing they are not so keen on, to their descriptions.
- They must not add the name of the game.

Share the descriptions on the class blog, or in the classroom.

Ask the learners to read the game descriptions.

- *How many names of games can the others guess?*

In the 'hood

Famous places near you

The learners use a geolocation app to find out about famous places in their neighbourhood.

Run up

Geolocation, or location-based, apps use your geographical location to show you information about places nearby.

There are a number of possible apps you can use for this activity, but here we suggest:
Wikitude – *http://www.wikitude.com/*

This pulls information from (and links to):
Wikipedia – *http://www.wikipedia.org/*

Try out the activity first yourself:
- Check what information Wikihood displays for your location.
- If none, try other geolocation apps and adapt the activity, depending on what places appear in the app.

The learners will need one geo-enabled device per pair for this activity.

Run

Ask the learners:
- *What do they know about this neighbourhood/town?*
- *What famous landmarks or places are nearby?*
- *What do they know about these places?*

Write any places the learners mention in a list on the board.

Put the learners into pairs, with one geo-enabled device per pair.

They open the Wikitude app, and and grant it permission to use their current location.
- This is vital for the app to work properly – it must know where you are, in order to provide useful local information.
- A list of nearby landmarks and places will then appear.

Ask the learners to check how many of the places on the board appear in the Wikitude list:
- *Are there any important omissions?*
- *Are there any places they don't know much about?*

Tell them to choose one of the lesser-known places in the list provided in Wikitude:
- They read further, by clicking on the name.
- This will take them to a Wikipedia article about the place.

Tell them to compile a 'fact sheet' of five little-known facts about their chosen place, in the form of a list. Help the pairs with language, as necessary.

Regroup the pairs, so that each person has a new partner:
- They share their five facts.
- They note down any new facts they learn from their partner.

The learners go back to their original pairs, and share with their partner any new facts they have learned.

Conduct class feedback:
- *What new things did they learn about this neighbourhood or town?*

Run on

Ask the learners to create a 'fact file' slideshow of images with subtitles for each of their facts:
- They upload the slideshows to a class blog.
- They share the slideshows in a subsequent lesson.

We revisit geolocation apps in Part C, and this activity is a good introduction to understanding how they work.

Chapter Four

Hands on!

Audio

In Chapter Four, the learners work mainly with audio. In keeping with our staged approach to working with devices, we introduce the use of audio after the learners have got to grips with working with text on devices in Chapter Two, and with images in Chapter Three.

Here, we combine all of these elements to provide the learners with fun and motivating audio-based mobile activities to produce and practise language.

The earlier activities in this chapter require the learners to use the built-in audio recording feature of mobile devices – which even very basic mobile phones have nowadays.

Audio recording may be a little more 'technically' challenging for learners who have never done it before, but they very quickly get the hang of it.

In fact, there is often more of a *linguistic* challenge involved in the activities where the learners need to record themselves speaking English, so we need to remember to give them plenty of language support, and enough rehearsal time.

The first three activities simply get the learners to listen to audio on their devices, and use this as a springboard for language work. We then start to move towards learners producing audio themselves.

The next three activities need the learners to produce clear, intelligible English – for example, in a 'voice search race' in *Inventions* and with an audio translation app in *Lost in translation* – and are especially effective for speaking/pronunciation work.

The rest of the activities in this chapter ask the learners to audio-record themselves and their classmates – in a range of different ways and talking about different things.

These activities also utilise a range of different audio recording app types, and combine text and images with audio in many cases.

Some activities can be done without an internet connection, but others do need connectivity, especially to be able to use web-based recording apps and to then share the learners' work online.

You can also share the learners' audio work by playing recordings back to the class and plugging speakers into the devices. If the recordings include visual elements, you can also connect the devices to a projector.

C Connectivity

This icon at the top of an activity indicates that at some point the learners will need to be connected to the internet.

M Mobility

This icon indicates that at some point in the activity the learners will be asked to 'go mobile'.

Musical me

My top five songs

The learners share their favourites.

Run up

In a previous lesson, check whether the learners keep music on their mobile devices. If not, ask them to add 10–15 songs they like to their devices before the next lesson.

Choose *your* favourite five songs from *your* mobile device:
- Put them into a playlist for quicker access during class.
- Include songs which have a special significance for you.

Run

Show the learners your mobile device:
- Tell them you are going to play them the first few seconds of your favourite five songs.
- As soon as they think they recognise the song, they should put up their hand, or call out *Stop!*

Play the first song for 10–30 seconds, or until one of the learners recognises it.

Elicit the name of the song and the singer/band.
- Explain why the song is significant to you, or why you especially like it.
- Add the titles and performers for each song to the board.

Put the learners into groups of four, to choose *their* 'top five' favourite songs.

They play the first few seconds to their group:
- The group members stop the song when they recognise it.
- They note down the song titles and singers/bands chosen by each member of the group.
- The learner playing the song explains to the group why they particularly like this song, and/or whether it has special significance.

Once the learners have noted down the top five songs for all their group members, ask them to negotiate a *group* top five. All the group members should note them down.

Put the learners into pairs from different groups, to share their group top five and to explain:
- *Whose choice was each individual song originally?*
- *Why it is significant to (or especially liked by) that person?*

Conduct class feedback:
- *Did any of the groups choose the same songs for their group top five?*
- *What kind of music was most popular in each group?*
- *Are there any interesting stories to share about the significance of any of the songs?*
- *Who chose the most currently popular or well-known music?*
- *Who chose the most unusual or little-known music?*

Run on

Ask the learners to record short samples of three or four pieces of music or songs that they hear outside class, in public places or at home (in shops, in a restaurant, on TV …) and to bring them to the next lesson:
- They share their recordings in groups.
- The others guess what the music is, and where it was probably recorded.

Conduct class feedback.

A short variation or extension to this activity is to play your usual mobile phone ringtone for the learners and tell them why you like it.

In groups of four:
- They play *their* ringtone for their group.
- They explain why they like it.

This can prepare them for the next activity: *Ringtone movie* on page 71.

Ringtone movie

Film soundtracks

The learners use a ringtone as the soundtrack
for an imagined movie.

Run up

Download an unusual or evocative ringtone sound effect
to your mobile phone. For example, from:
www.mediacollege.com/downloads/sound-effects

Choose instrumental or 'sound effect' music, rather than
a song with words:
- *A whistling cowboy tune*
- *Scary-sounding music*
- *A lively jazz track …*

Run

Tell the learners you are going to play them the soundtrack
to an imagined movie:
- They should imagine a movie screen in front of them.
- They should close their eyes and listen to the soundtrack.
- They should imagine what's happening in the movie –
  from the soundtrack:
  The location
  The weather
  The characters
  The scenery
  The action
  The dialogue

Play the ringtone for a minute or so, giving the learners time
to sit and listen with their eyes closed.

Stop the ringtone, and ask the learners to write down what
they saw in their imagined movies – in as much detail
as possible:
- Get them to include a description of the scenery and
  characters, as well as what was happening on the screen.
- Get them to write freely for 5–10 minutes, to 'capture'
  their movies.

Play the ringtone again:
- The learners read through their notes.
- They make any changes that they think are necessary
  to fit the music.

Allow them another 5–10 minutes for reviewing their texts
and making corrections. Help with language, as necessary.

Put the learners into groups of three or four:
- They 'tell' their movies – without just 'reading aloud'.

In their groups:
- They negotiate a combination of parts of each story into a
  final group story, to present to the rest of the class.

Give each group a chance to read their story out to the class,
accompanied by the soundtrack.

Conduct class feedback:
- *Were any of the movie stories or scenes similar?*
- *What would the ideal final story be?*

Run on

Ask the learners to choose and download an unusual
ringtone to their phones:
- Put the learners into groups of three.
- Tell them to combine their three ringtones into one story.

The learners play their ringtones and tell their stories to the
rest of class.

Acknowledgements: Thank you Lindsay Clandfield for the
original idea for this activity.

Daily noises

The sounds of my day

The learners record sounds from their daily routine.

Run up

Record on your mobile device five different sounds from
your daily routine, each recording lasting about 10 seconds:

- *Radio or TV*
- *Shower running*
- *Kettle whistling*
- *Train or bus*
- *School bell ringing*
- *Children playing*
- *Clock ticking*
- *Dinner cooking …*

Ask the learners to do the same, and to bring their
recordings to class on their devices.

Run

Explain that you are going to play the series of sounds
recorded during your typical day:

- The learners listen and note down the sounds they can
  identify.
- You allow them to check their notes in pairs.

Play the sounds again, if necessary.

Elicit the sounds onto the board. For each of the sounds,
ask the learners:

- *What was I doing?*

Elicit past continuous sentences in response:

- *You were having a shower.*
- *You were making tea.*
- *You were catching a train …*

Alternatively, elicit the question form:

- *Were you …?*

For lower levels, use the present continuous tense,
highlighting the language if necessary:

- *to be + …ing*

Put the learners into pairs, to listen to each other's
recordings:

- They create past (or present) continuous sentences,
  based on what they hear.
- Alternatively, they ask each other questions.

For each sound, each learner should note down what their
partner was doing.

Conduct class feedback:

- *What were most people doing?*
- *Who was eating?*
- *Who was catching the train/bus?*
- *Who was watching TV?*

Run on

Ask the learners to write a short paragraph about their
partner's daily routine, based on the sounds they heard.

Highlight that the present simple tense should be used here
– elicit the first few lines of your own routine onto the board
as an example:

- *Gavin has a shower, and then makes a cup of tea. He catches
  the train to school …*

Encourage the learners to add their paragraphs to a class blog
or portfolio. They then visit their partner's description of
their daily routine, and comment on one thing they *also* do
every day, in the 'Comments' section.

Inventions

A voice search race

The learners find out about famous inventions, using a voice search engine.

Run up

Prepare a quiz on a handout – with about eight inventions:

What?	Who?	When?
Atomic bomb		
Bifocal glasses		
Car		
Light bulb		
Plastic		
Printing press		
Telephone		
X-ray		

Run

Show pictures of the inventions from the quiz (or write them on the board):

- *What are they?*
- *Which inventions have helped mankind most?*
- *Which have not?*
- *Why?*

Distribute your handout.

Put the learners into pairs, and ask them to fill in the information they already know:

- *Who invented each?*
- *When?*

In their pairs, they write down questions for the information they don't know. Elicit one question onto the board as an example:

- *Who invented the car?*
- *When was the car invented?*

Highlight the 'active versus passive' question forms, if needed.

Check all the learners' questions, paying special attention to pronunciation:

- They will need to use a voice search app to find the information they don't know.
- Their questions need to be intelligible!

Tell the learners to open Google Voice Search on their devices:

- Tell them to ask their questions, and to note down the answers on the handout.
- Tell them this is a 'search race', so the first pair to find all the answers and fill in the handout must shout *Stop!*

When the first pair finishes, stop the race and check the answers with the class:

- Each correct answer is awarded one point.
- Check how many points each pair has scored.

Each pair now writes five *new* questions about *different* inventions, and notes down the answers separately. They have to search for the information and the answers (if they don't already know them) using Google Voice Search.

Ask the pairs to exchange questions.

Set a time limit of five minutes for the pairs to find the answers, then stop the activity:

- The learners give their questions back to the pair who wrote them.
- They check whether their own questions were answered correctly, and give feedback.

Conduct whole-class feedback:

- *Which questions were the most difficult?*
- *Which were the easiest?*
- *Which were the most unusual or surprising?*
- *What is the most interesting fact they have learned?*

Run on

Show the learners images of *unusual* inventions. For example, from:
http://goo.gl/1Mt7ph

Put the learners into pairs and ask them to choose one invention, and to write a 100-word 'biography' of the invention, describing *what* it is, *who* invented it, and *when*.

- They upload their invention to a class blog or portfolio.
- They include an image (or provide a link to the image if it is not 'Creative Commons' licensed).

They read about the other inventions, and leave comments.

In a subsequent lesson, show the inventions and biographies with a projector:

- *Which invention do they like the most?*
- *Which is the silliest?*
- *Which is the most practical?*
- *Which would they buy?*

You can also leave a comment yourself on their inventions and biographies in the class blog or portfolio.

Listen to me!

Pronunciation practice

The learners' speech is converted to text.

Run up

Download a speech to text/dictation app and try it out, to ensure you know how it works. You could practise, using the script opposite.

Ask the learners to download a dictation app to their devices before the lesson.

Run

Ensure that your dictation app is working and projecting onto the screen in class.

Speak into the microphone of your device:
- Tell your learners a little bit about the app.
- Explain how it can be used for pronunciation and fluency practice.
- You can use the script opposite.

Check through the transcription with the learners, and highlight any errors you (or the app!) made:
- *Why do they think these elements might have been problematic?*
- *How do they think an app like this might help them?*

Distribute short texts for your learners to read into the dictation app on their devices.

Try to concentrate on individual areas of difficulty within the texts and, if possible, give the learners:
- Texts that you know will present some challenges.
- Texts that don't present *so* many difficulties that the learners are *too* challenged – and find it dispiriting.

Give the learners time to read their texts, working in pairs and helping each other if they run into problems.

The main motivator in an activity of this nature is for the learners to achieve as close to 100% recognition as possible:
- Allow them all the time they need.
- Allow them as many attempts as they want.

Monitor each pair, helping them with problem areas.

Conduct feedback as a whole class:
- *Did they find this tool useful?*
- *Will they continue to use it at home or outside class?*

Speech to text

Today we're going to look at how we can get speaking practice, using apps on mobile devices.

If you look at my screen, you'll see that what I'm saying is appearing on the screen – as text.

This is an amazing app.

Here's how it works:
- *When I speak, my audio is sent over the internet to a server.*
- *The server converts the audio into text, and sends the text back to me.*
- *I can then take that text and use it – in emails or documents, for example.*

To make this work, you will need to speak clearly and carefully. This can help you with your pronunciation and fluency.

Are you ready to try it out?

Run on

- In monolingual groups, focus on individual problem areas on a regular basis – by giving the learners a text to experiment with at home.
- In multilingual groups, consider preparing individualised texts for each learner – time permitting.

Dictation apps can be especially effective for individualised work on pronunciation for learners:
- It is worth encouraging them to use the app regularly out of class.
- You can set short texts or extracts from the coursebook, for example, for the learners to read into the app for homework.

This will also encourage them to review the course content at the same time as practising their pronunciation.

Lost in translation

Deconstructing conversations

The learners reconstruct bad translations.

Run up

Download and install an audio translation app to your device.

Create your own sample translation, using a short original English text. Dictate the text into the audio translation app, and convert it to the learners' first language.

- If you're working in a multilingual environment, you will need to translate your English text into each language present in class (see the steps below).
- If you have very advanced groups, running the translation through a few other language translations will obscure the original text even more, and this will make for a more challenging activity.

Ask your learners to install an audio translation app on their devices.

Our example opposite uses the audio translation feature of the Google Translate app.

Run

Start with a short discussion with the learners on attitudes to dictionaries, phrase books, translators, etc:

- *How often do they use a dictionary in class or at home?*
- *Do they prefer a paper or an electronic dictionary?*
- *Have they ever bought a phrase book?*
- *Do they use translation in their English learning?*
- *Have they got any translation apps on their devices?*
- *Do they often use them?*
- *What do they think are the potential problems with these kinds of tools?*

Give out (or project on screen) the translated version(s) of your original text.

The learners look at the translation, and try to write what they think the original English version was.

- Note that the translation into Spanish opposite is *not* accurate, and proficient Spanish speakers will immediately see the pitfalls of automatic translation in this example.

Feed back as a whole class, working through the suggested changes until you get to a correct text which is the same as – or very close to – the original.

Ask the learners to jot down a short text about themselves in English:

- They record a translation, using Google Translate.
- They exchange it with a partner.
- They reconstruct the texts in correct English.

Spanish translation – what the app heard Gavin say.

- Mi nombre es Gavin y yo soy un entrenador maestro profesor y un escritor. Escribo libros sobre tecnologías de la educación y la aplicación en la formación docente. En mi tiempo libre, me gusta salir a comer tomar vacaciones largas de conducción y escuchar música. También toco la batería y el ukulele uno no muy bueno ninguno de los dos.

Original version – what Gavin said.

- My name is Gavin and I'm a teacher, teacher trainer and a writer. I write books about educational technologies and their application in teaching training. In my spare time, I like to go out to eat, take long driving holidays and listen to music. I also play the drums and the ukulele, though I'm not very good at either.

Feed back as a whole class, and discuss the errors produced and the perils of automatic translation.

With multinational groups, compare the quality of the translations produced:

- *Which languages translated better?*

Run on

Translation is enjoying a new lease of life in English language teaching, and there are plenty of good activities that can be done using it:

- Encourage the learners to look for examples of bad translations when they are out and about.
- Restaurant menus are generally good for these, as are product labels, manuals and shop window notices.

Ask the learners to take photographs of these examples and bring them to class for more language work:
See *All around me* on page 65 for more on this.

Philip Kerr has some excellent resources and suggestions for using translation on his website:
http://translationhandout.wordpress.com/about/

The place I love

Animated audio guides

The learners describe their favourite place.

Run up

Create a recording of yourself describing a photograph of your favourite place.

To do this, use an animated movie app like Tellagami to create an avatar that looks a bit like you, with an audio recording to describe what can be seen in the photo.

Alternatively, you can use a presentation app – like myBrainshark – that allows audio voiceover.

Try to include answers to the questions in the activity below. You can see an example of the kind of thing to aim for here: *https://tellagami.com/gami/J4WE9L/*

Also make sure that the learners have an animated movie app or presentation app installed.

Run

Ask the learners to think of a place they like:
- *Where is it?*
- *What is it like?*
- *When do they go there?*
- *When was the last time they went there?*
- *What do they do there?*
- *Why do they like it?*

Show them the background picture from *your* example 'place I love':
- Get them to brainstorm questions about it.
- Get the questions up on the screen or board.

Play your recorded example.

Feed back as a whole class:
- *Did they get answers to their questions?*

Answer any unanswered questions.

Play your example again – this time asking the learners to note down what you talked about.

Work together to produce a template for the ideal script.
- Note that if you use Tellagami for this activity, you can only record 30 seconds of audio in the free version, at the time of writing.

Ask the learners to script their descriptions, and then put them into pairs to do a round of peer correction.

Briefly show them how to create their favourite place recordings. For example, if you use the Tellagami app, show how to do the following:
- Add a background.
- Customise their avatars.
- Record their voice.

Working individually, they locate an image of their favourite place, either by finding it on their devices or by downloading one from the internet.

They create their guides, following the script they wrote.

Encourage the learners to share their recordings by adding the URLs (or the finished products, depending on the app they used) to a class blog or wiki.

Run on

There are many different things you can do with these recordings in subsequent lessons. For example:
- Language work, based on the scripts the learners have recorded.
- A virtual tour of favourite places, viewing the recordings as a class with a screen and projector.
- Voting for the most unusual or interesting-sounding places.

Speaking through avatars:
- Gives the learners time to organise their thoughts before 'speaking', and to rehearse.
- Provides valuable pronunciation and fluency practice.

They can also be very beneficial to learners who struggle with confidence issues and anxiety connected to speaking in public.

This is my life

Audio interviews

The learners conduct interviews with you, with staff and with friends.

Run up

You may want to prepare some possible answers to the questions you think your learners might ask you.

Run

Tell the learners they are going to interview *you* about *your* life outside the school.

Put them into small groups (A, B, C, D …) to brainstorm:
- *What kinds of questions would they like to ask?*

You may want to give them some ideas, to help them with the questions:
- *Hobbies and interests*
- *Weekend activities*
- *Holidays*
- *Family*
- *Daily routine*
- *Past life (school, other jobs)*

Combine two groups into bigger ones (Group A with Group B, Group C with Group D …) and tell them they can ask a maximum of 10 questions.

Give them time to negotiate from their lists.

Repeat this process until the whole class has agreed on just 10 questions.

The learners choose a class interviewer:
- They use their audio recording application on their mobile device.
- They conduct the interview.

Conduct group feedback:
- *Are they surprised at anything they found out?*

The recorded interview can be uploaded to a class blog or wiki, along with a photo of you.

Now put the learners back into their original groups, to choose someone else who works in the school (the principal, another teacher – it must be someone who speaks English).
- They brainstorm possible questions – this time about that person and their job.
- You arrange for them to meet with the people – and interview them.

When they have finished their interview, ask the learners to transcribe it and write it up:
- This can be in the form of a poster.
- Alternatively, it can be a blog or wiki entry with the audio files, accompanying photographs – and more.

Run on

Ask the learners to read or listen to the interviews for homework:
- *Who has the most interesting job?*

In the next lesson, distribute the transcript from one group to another group:
- They review the language.
- You then conduct class feedback and error correction.

If you are in a school in an English-speaking country, or have access to other English speakers outside your immediate place of work:
- The learners can interview someone outside the school (their host family, for example).
- They bring the interview to a future lesson, along with some photos.

Conduct a 'show and tell' with each of the interviews.

What is it?

Object clues

The learners draw and describe common objects.

Run up

Prepare a short recording, describing an everyday object (see the example opposite).

Run

Tell the learners they are going to listen to a description of a normal, everyday object:

- Make sure that it's an object they are all familiar with.
- Try to be as vague as possible in your description, to make it more challenging.

As they listen, they should draw the object in a drawing app on their mobile device. For an example, see opposite.

Ask the learners to put their device on their desk, with the sketch visible and walk around the classroom to see what everyone else has drawn.

Conduct feedback as a whole class:

- *Which clues helped them guess the object?*

Before continuing, you may need to do some language work on describing objects:

- *Materials*
- *Colours*
- *Adjective order*
- *Purpose …*

The learners secretly identify an object in the class, and write a short description of it:

- They give clues.
- They avoid being too obvious.

When they are happy with their descriptions, they record them using an audio recording app on their phone.

Play each recording for the whole class:

- *Can they identify each of the objects described?*

Make some notes on any language in the recordings that needs work, and conduct whole-class feedback and correction on the board or a connected screen.

Run on

The same activity can be done out of class:

- The learners describe objects they encounter in town or at home.
- They bring these recordings to class.

A much more sophisticated version of this activity – called Invader – is suggested by Paul Driver: *http://digitaldebris.info/invader/*

'This is something you find in most offices. It is usually made of metal or plastic, though they used to be made of wood, and you usually see it with the pens, pencils and other stationery items. You get short ones and long ones. People generally use them for drawing straight lines, but they're also good for opening letters.'

Answer:

Talking trash

Stories from the street

The learners take a photo of rubbish, and record the story of how it got there.

Run up

In the week before the lesson, ask the learners to look out for two or three pieces of rubbish/trash in the street, photograph them on their mobile devices, and bring their photos to class – the more interesting or unusual, the better.

Take a photo of a piece of trash yourself.

Prepare a story about *your* piece of trash, and how it got there, and write it on a handout.

Run

Show the learners your photo of what you found in the street:
- *What do they think it is?*

Tell them that you are going to tell them the story of this piece of rubbish.

Tell the learners your story:
- Use your own mobile device to audio-record yourself telling your story.
- Opposite is an example story, based on a photograph of a soft drink can lying in the street.

Make sure that the learners have understood the story, by asking brief questions. For example:
- *Where was I born?*
- *Who bought me from the supermarket?*

Play your audio-recorded story again to the learners, if necessary, to help them remember the details.

Give out a written copy of the story, and ask the learners to identify the tenses (*simple past*; *simple present*) and which person it is written in (*first person – I*).

Tell them what they are going to do:
- In pairs, compare their trash photos on their devices.
- Choose one photo to create a story for.
- Prepare the story outline together.
- Audio-record their stories on their mobile devices.

Give the learners a clear time frame in which to prepare and rehearse their stories (10–20 minutes, depending on their level of proficiency):
- Help them with language and pronunciation.
- Ensure that they use a first person narrative perspective (*I*, not *it*).
- Ask them to send you their chosen photos of trash, eg via email or a messaging program.

Collate all the photos into one slide or upload them to a Flickr account, for the next stage of the activity.

I was born in a factory. I spent the first few months of my life on the shelf of a supermarket. It was quite interesting watching people coming and going every day, but after a few months I felt like nobody really noticed me.

One day a boy took me off the shelf, paid for me and took me home. The next day he took me to school, and drank me during the lunch break. Then he and his friends kicked me around like a football as a game.

That's how I got these dents and scratches. One boy kicked me so hard I flew over the school wall, and landed in the street. A car ran over me.

That's why I'm so flat. But it's nice living here on the street. I watch people, cars and dogs pass by all day …

For the recording phase:
- If possible, send the pairs to different quiet areas in the school.
- If not, move them as far away from each other as possible.

Both members of the pair should contribute equally to the recording.

In class, show the learners the photos you collated while they were working on their recordings, and play each audio story:
- *Which photo is each story describing?*

Alternatively, ask the learners to upload their audio stories and photos to a class blog for homework, and to listen to each other's stories and leave a text comment on each story.

Run on

Ask each pair to exchange stories with another pair:
- They record a continuation of the story.
- They explain what happens next to the piece of rubbish.

Conduct feedback by playing some of the audio recordings to the whole class.

Acknowledgement: Thank you Shelly Terrell for this idea.

Crazy sports

The rules of the game

The learners invent a sport and create a screencast describing the rules.

Run up

Choose a sport for which the learners know the rules.
- Check the rules yourself, if necessary.
- Note these down, using modal verbs of obligation.

Run

Elicit team sports from the learners, and write them on the board:
- *Football*
- *Waterpolo*
- *Rowing …*

Elicit a few rules for each sport – to review modal verbs of obligation:
- *must, mustn't, need to, have to …*

Tell the learners you are going to describe the rules of one sport – without telling them the name:
- *Can they guess what it is?*

See opposite for an example sport.

While you describe the rules:
- Draw the field and players on the board.
- Use a drawing app with voice recorder, if you have a tablet computer connected to a projector.

Give each learner a slip of paper with a sport.

Put them into pairs and tell them they are going to invent the rules of a *new* game, which is a combination of the *two* sports on their slips of paper:
- Basketball + golf
- Swimming + football
- Tennis + ice hockey …

Ask the learners to give their new sport a name:
- Basketball + golf could be *baskgolf* or *golfbask* …
- Swimming + football could be *swimball* …

They now have to invent the *rules* of their new sport:
- Give them some time to rehearse describing the rules.
- Encourage them to use modal verbs of obligation.

When the learners are ready, ask them to draw the field, the positions of the players, etc, on a tablet – while describing the rules in a screencast app.

Regroup them, to watch their classmates' descriptions of the new sports.

Close the lesson by asking:
- *Which sport(s) would be the most difficult to play?*
- *Which sport(s) would be the easiest to play?*

- *You need two teams and a field.*
- *You must have 13 players in each team.*
- *There are posts at each end of the field, shaped like the letter H.*
- *Each team needs to defend their half of the field.*
- *The ball is oval.*
- *The players have to work together to carry or kick the ball to the other end of the field, and place it over the line.*
- *You can physically stop the person with the ball, but you mustn't try to stop someone without the ball.*
- *When a team member gets the ball to the other end and places it over the line, the team gets points.*
- *The team must then try to get extra points by kicking the ball over the H-shaped posts.*
- *The team has to try to score as many points as possible.*
- *The team with the most points wins.*

Sport: rugby league

For more on this sport, see:
http://en.wikipedia.org/wiki/Rugby_league

- *Which is the most dangerous? Which is the least dangerous? Why?*
- *Which sport would they most like to play? Why?*
- *Which sport would they least like to play? Why?*

Run on

Ask the learners to upload their screencast recordings to a class blog or portfolio:
- They visit their classmates' screencasts.
- They leave a comment on each.

Round up the activity by leaving a comment yourself about each of the recordings.

Make sure you accentuate the positive – don't only provide feedback on language errors!

Acknowledgements: Thank you Alicia Artusi for the screencast idea for describing how sports are played, and Mariana Mañueco for the idea of combining two sports to create a new game.

Follow me!

An audio treasure hunt

The learners follow audio instructions
as a series of clues.

Run up

Prepare a set of audio clues, using an audio recording app
(the example below uses Audioboo).

Ensure that your learners have an audio recording app
installed.

Note:

- All your clues will have long URLs. For example:
  http://audioboo.fm/boos/1690403-clue-1
  You might want to shorten them, using something like
  Google URL Shortener, which gives much shorter links:
  http://goo.gl
- To do this, paste your long link into the box and click on
  the 'Shorten URL' button to get a short link. The audio file
  above converts to:
  http://goo.gl/K6jx7L
  This is then much easier for your learners to type into
  their browsers.

Print the short URLs of your audio files and put them in
strategic places, in the school to create a treasure hunt.

Run

Tell the learners that they are going to go on a treasure hunt
around the school – all the clues are audio clues:

- They can listen to the clues by using the short URLs
  at each destination.
- The clues all give an instruction, which they should follow.

Review the language for giving directions, if necessary:

- *Turn left/right.*
- *Go straight ahead.*
- *Head straight on …*
- *Keep to the left/right.*
- *Double back.*

You can use language that is more, or less, complex for
giving directions – depending on the level of the learners.
With beginners or elementary learners, you can keep the
language in the audio clues very simple indeed!

Tell them that the first clue is on the wall in the classroom:

- All they need to do is type the short address into their
  web browser. For example:
  http://goo.gl/K6jx7L
- They will be taken directly to the audio file:
  *'Go out of the class and turn right. Head straight on through
  Reception to the Canteen and find the next clue by the coffee
  machine.'*

Make sure that your last clue brings them back to class!

In pairs, send the learners off on the treasure hunt.

- You might want to have them collect objects, as they
  follow and find the clues.
- You might ask them to take a photo of each place.
- You might offer a prize at the end.

Give the learners time to complete the treasure hunt,
then conduct feedback as a whole class.

Now show them:

- How to set up an account at Audioboo.
- How to create their first recording.
- How to create the short *goo.gl* link for it.

Put them into small groups, to plan their own short
treasure hunt around the school.

Once they have their treasure hunt ready, the groups
exchange 'start points' and complete each other's hunts.

Run on

Why don't you get your learners to prepare a treasure hunt
for *you* to do?

These treasure hunts become even more motivating when
they take the action outside the school and into the local
town or city.

This activity combines well with the use of QR codes as
prompts to the audio files.

The following code will take you to our example clue:

To read the code, you will need a QR code reader app.
The camera on your device will read the code and open
the linked resource – in this case, the audio file.

We will be looking at QR codes in detail later in the book –
see the 'QR codes' section in Part C on page 97.

Chapter Five

Hands on!

Video

In Chapter Five, we continue to build on the skills learners have developed in the previous chapters. By now, they have worked with text, image and audio. The activities here focus on video, or on a combination of text, image, audio and video.

As with audio recorders, even the most basic mobile phones tend to have video recording capabilities these days, so several of the activities in this chapter can be carried out on less sophisticated phones.

Other activities require an internet connection, especially if you would like your learners to share the videos they create online.

Video recording and editing is arguably the most technically challenging for learners who have never done it before. We recommend apps that can simplify video editing and production, and that give their videos a professional look and feel.

As with recording audio, there is often more of a *linguistic* challenge involved, especially at lower levels. So this is where teachers need to focus most of their attention during class – in giving learners the language support they need.

The use of technology is always secondary to this fundamental aim.

The activities in this chapter involve the learners mainly in producing video themselves, rather than just passively watching.

Although some learners may regularly record video on their mobile phones in their personal lives, they are probably less familiar with recording video in an English class – so they may feel more comfortable filming objects or places, rather than themselves, when they start creating videos in English.

The first four activities in this chapter get the learners to film objects, words and places without a narrated audio soundtrack, or to create subtitles. This means less pressure on them while they are getting used to creating video in and for class.

In the rest of the activities, the learners provide an audio narrative for videos that they create. The first of these (*It's mine!*) provides a teacher model for the learners to work from.

The activities cover a wide range of topics – including biographies, news, holidays, music, hobbies, movies and books.

In each case, the learners produce videos using a range of apps, and – most importantly – they use the videos for language work.

C Connectivity

This icon at the top of an activity indicates that at some point the learners will need to be connected to the internet.

M Mobility

This icon indicates that at some point in the activity the learners will be asked to 'go mobile'.

Silent movies

House tours

The learners create a silent video tour.

Run up

Ask the learners to create a two-minute silent video 'tour' of their house/flat and bring it to class.

Do the same yourself, and prepare 5–10 questions about your video (see below for some examples).

Run

In class, show your movie tour, describing the rooms and furniture as they appear on the screen. Ensure that you describe the items you will be asking questions about later:
- The learners listen and watch closely.
- They try to remember as much detail as possible.

Ask the learners questions, to see how much detail they remember from your silent movie. For example:
- *What is the second room in the video?*
- *What is the last room?*
- *What colour is the tablecloth on the dining room table?*
- *How many cushions are there on the sofa?*
- *What is on the living room table?*

Ask them to individually review *their own* silent movies.
- Learners at lower levels note down the key words they will need to describe their movies aloud.

Ask them to prepare five questions about the rooms and items in their videos.
- Learners who haven't brought a movie to class are paired with a learner who has – they will be working together.

In pairs, the learners show each other their movies, while providing a running commentary:
- They ask their partners the questions they prepared.
- They award a point for every question answered correctly.

Round up by asking:
- *Did anyone manage to score five points, by answering all five of their partner's questions correctly?*

Run on

Ask the learners to record their commentaries over their silent movies:
- They use an audio voiceover app.
- They upload the finished videos to a class blog or wiki page.

In this way, they can view and comment on each other's videos.

Acknowledgement: Thank you Lindsay Clandfield for the idea for this activity.

On the vine

A six-second video

The learners create very short videos on anything they like.

Run up

Explore Vine (*https://vine.co/*) – a video collage app that enables you to create six-second videos with different scenes stitched together and then played in a loop.
- Note that Vine requires users to be over 17 years of age, so for younger learners, use a different video collage app.

Run

Ask the learners if they know Vine.

Using the class projector connected to your mobile device, show them four or five recent Vines from the app homepage.

Elicit the learners' reactions by asking:
- *Which Vine videos did they like best? Why?*
- *What makes an effective Vine video?*
- *What makes some of these videos so creative?*

Tell the learners they will each record a Vine in the school:
- In pairs, give them five minutes to think about and discuss possible locations and subjects. (Giving the learners free rein in their choice of topic/location will encourage them to be much more creative.)
- In pairs or individually, tell them they can now record their Vine videos.

Give them a clear time limit (eg 10 minutes) to go around the school and to create their six-second Vines.

Back in class, connect the learners' devices to the projector and watch the Vines.

After showing each one:
- Ask the learners to explain why they chose that particular subject matter.
- Ask the rest of the class for feedback on the video.

Run on

Ask the learners to each create a new Vine outside of class – either in the street or at home.

In the next lesson, show some of the videos and ask for class feedback:
- *Which places did they choose for their new Vines?*
- *What 'mood' did each of these Vines create?*

Ask them:
- *Did they enjoy the creative process behind making these very short videos?*

If they did, invite them to make more Vines in the future.

Visual poems

Stop motion videos

The learners create a poem,
using stop motion video techniques.

Run up

Ensure that the learners have a stop motion app on their devices, and that there is one device per pair of learners.

Familiarise yourself with the app by creating your own stop motion poem. For example: on 'Summer'.

One of the simplest ways to make a stop motion video is to photograph words/sentences on paper.

Run

Put a number of evocative nouns on the board. Choose words that could work well as the topic of a poem:

- Love
- Loneliness
- Spring
- Winter
- Forest
- Desert
- Mother
- Child
- Sea
- Mountain
- Speed
- Tranquillity …

In pairs, the learners brainstorm two or three words for each topic.
- Elicit the words.
- Add them to the board.

Write up another topic (eg *Summer*) – and construct a poem on the board with the whole class, by adding the elements in the example opposite.

Tell the learners they are going to create a poem in pairs:
- They choose one of the topics for the subject of their poem.
- They write their poem.
- They follow the structure of the 'summer' poem.

Show them your own stop motion version of the poem.

Now ask them to create a visual poem, using a stop motion app:
- First ask them to write each line of their poem on a separate piece of paper (they could draw an image to accompany each line if they wish).
- Then ask them to use the stop motion app to take a photo of each line.

Finally, show the finished visual poems to the class by connecting the learners' devices to a projector.

Summer

Summer. (the topic)

A warm summer night. (the topic plus two adjectives)

The smell of the sea. (a smell)

Watching the stars in the sky. (an action)

Feeling the sand on my toes. (a sensation or feeling)

On the beach. (a place)

Summer. (repeat the topic word)

Run on

Ask the learners to create a second poem on any topic, following the same structure as before:

Topic
Topic plus two adjectives
Smell
Action
Sensation or feeling
Place
Topic

They then film each line of their new poem with the stop motion app.

The finished poems can be uploaded and shared in a class blog.

Bollywood

Subtitled Hindi movies

The learners create subtitles
for Bollywood film clips.

Run up

Explore BombayTV and choose a film clip to subtitle
with the class. See:
http://www.grapheine.com/bombaytv/

Note that BombayTV requires Flash, so it is not suitable for
iOS devices at present – use another device (eg Android) or
a laptop, to demonstrate this activity with the learners.

Run

Write *Bollywood* on the board and ask:
- *What is it?*

Give the learners a two-minute time limit to find out as
much as they can about Bollywood on the internet, using
their devices.

Tell them they are going to subtitle an original Bollywood
film clip.

Show the film clip you chose from BombayTV, using the
class projector, then elicit possible contexts and scenarios:
- *Who are the people in the clip?*
- *What is their relationship?*
- *What is the situation?*
- *What is the atmosphere like?*
- *Is the dialogue friendly, joking, angry, passionate, sad …?*

Once the learners have established a possible context, elicit
ideas for the dialogue and add the subtitles to the clip,
showing them how the site works.

Save the finished product, and play it through a couple of
times for the learners:
- *Are they happy with the length of the characters' lines?*
- *Does the dialogue fit the scenario?*

Put the learners into pairs:
- They choose a different clip from BombayTV.
- They establish who the characters are, the context,
  the tone of the dialogue, etc.
- They note all this down – they will need to present this
  context to the class when they share their subtitled video.
- They subtitle it, using non-iOS devices or laptops.

When the videos are ready, ask each pair to present theirs
to the class, via the classroom projector. They first need
to describe:
- The context
- The relationship between the characters …

Round up by asking the learners:
- *Which films did they think were the most original,
  the funniest, the saddest, the most surreal …?*

Run on

The BombayTV site also offers the option to subtitle
B-movies (low-budget films generally of inferior quality).
- In 'Change the channel', choose 'Bmovie-TV' from the
  drop-down menu.
- The learners, individually or in pairs, choose a B-movie
  clip to subtitle.

They follow the same procedure as above, to create and share
the B-movies.

Acknowledgement: Thank you Joshua Davies for showing
us BombayTV.

It's mine!

Significant objects

The learners create videos describing an important personal object.

Run up

Bring a small significant personal object of yours to class:

- *A piece of jewellery or clothing*
- *A keepsake or souvenir*
- *A drawing*
- *A book*
- *A shell*
- *A toy …*

Ask the learners to bring a small significant personal object.

Run

Seat the learners around you, and show them your object.

Ask them to use their mobile device to video-record you as you describe your object – in just 3–4 sentences – and why it is important to you.

The learners work in pairs to review their video recording of you describing your personal object:

- They transcribe what you said.
- They add it to their video recordings, using a video subtitling app.

Play back one subtitled video recording to the class from one of the leaner's devices connected to a projector, so they can check their work.

Ask the learners to take out *their* significant objects. Give them five minutes to prepare 3–4 sentences:

- They describe what their object is.
- They describe why it is important to them, making notes if they wish.

Put them into pairs, and ask them to find a quiet space for recording the video.

Tell each pair to video-record each other describing the objects, using their partner's device. If they prefer, they can focus their cameras on just the object, rather than on each other's faces.

Run on

Put the learners into pairs, to exchange devices and watch their partner's recording:

- They provide a transcription for their partner's descriptions.
- They add it as subtitles to the video recording, using a video subtitling app.
- They upload the finished video with subtitles to a class blog or wiki page.

Conduct class feedback, asking each pair to show their partner's personal object to the class, and to tell the others about it and why it is significant.

A life in film

Video biographies

The learners create a biography of a favourite artist, actor, musician …

Run up

Make your own video biography to show the learners, or give them a sample of a student-created biography. Here is an example of what to aim for: *http://goo.gl/Rzr233*

Ensure that the learners have installed a video image app which allows them to combine voice and images.

Run

Put the name of the person from your own video biography on the board, and ask the learners to brainstorm in small groups:

- *What do they definitely know about the person?*
- *What do they think they know about the person?*
- *What would they like to know about the person?*

Conduct feedback as a whole class, then show your video:

- *Was what they knew correct?*
- *Did they find answers to their questions?*

If there are unanswered questions, try to answer them – or google the answers immediately, if you're not sure.

Ask each learner to choose someone they would like to make a video biography of.

Put them into pairs, to brainstorm what they know about their chosen person.

- This can be more productive if they interview each other.
- They can use sources like Google to check information and gather images.

Once they have gathered all the information and media they need, give them time to work on the script and the organisation of the video:

- This can be done on paper.
- It can be done in a drawing app.

The learners rehearse their biography with their partner, making sure they are ready for the recording phase, which will be done for homework.

Run on

The video production side of this activity is done at home:

- The learners prepare their video biography.
- They then share it – via YouTube or on the class blog.

Encourage the learners to explore the different biographies – and leave feedback, comments or questions.

Acknowledgement: Our thanks to Anglo European School of English in Bournemouth, UK, for allowing us to use their student-created movie as an example for this activity.

It's news to me

The students as reporters

The learners script a news bulletin.

Run up

Make sure you watch or read the news on the day of the lesson.

You may want to give your learners an example of a student-created news bulletin. Here is an example of what to aim for: *http://goo.gl/wukb3W*

Run

Start by telling the learners of an interesting news story from the day:

- It can be something you saw on TV or read online or in a newspaper.
- If you want to try something a little more interesting than the usual headlines, try a story from the Yahoo Odd News section:
  http://uk.news.yahoo.com/oddly-enough/

Put the learners into six groups.

They write down any news they can remember from the last couple of days, in one of the following categories:

- Local news (Group A)
- National news (Group B)
- International news (Group C)
- Entertainment (Group D)
- Technology (Group E)
- Sport (Group F)

Regroup as a whole class:

- Write the main topics on the board or a connected screen.
- Vote on one story for each category – these six stories will make up the news bulletin.

Put the learners back into their original groups, and ask them to script their story.

They do the following:

- They write the main story.
- They decide on the definitive headline.
- They decide if they need any graphics or images to accompany the story.

Once they have the script for the story, ask them to film their segment using one of their mobile devices.

- You can enhance this process with a suitable backdrop, or by displaying PowerPoint (or Keynote) slides behind the newsreader while they are reading their story.
- You can add suitable graphics and titles in post-production later – either on a mobile device or on a laptop or desktop computer, using a video production app.

When all the pieces have been filmed:

- If you have time, you can edit the final news bulletin as a whole class.
- If you are short of class time you can do it yourself at home, to show in the next lesson.
- If you have a class blog or wiki, you can share the news bulletin online.

Run on

This activity can be extended outside the class:

- The learners can go out into the town or city.
- They film a local story of interest to them.

They then share it with the class in the next lesson.

Acknowledgement: Our thanks to Anglo European School of English in Bournemouth, UK, for allowing us to use their student-created movie as an example for this activity.

Come with me

Video brochures

The learners create a video guide to their town.

Run up

Find a couple of video guides online. To show to your learners as examples, you can find student-created videos here:
http://goo.gl/gd5lZ9
http://goo.gl/CPfkw7

Ensure that the learners have a video production app on their mobile devices.

Run

Tell the learners the name of a place for which you have a video guide.
- They brainstorm what they know about the place.
- If you have connected devices, let them browse a few sites and read up about them in Wikipedia or on other sites.

Conduct feedback to the board or screen, noting the main points the learners have come up with.

Tell them they are going to watch a video guide for the place:
- *What would they expect to see in it?*

Write up their points before you play the video.

Play the video the first time, with the sound down.

Ask the learners what they saw:
- *How much did they guess right?*

Play the video a second time – this time, with the sound up – to compare the video with their original expectations:
- *What was in there, what wasn't?*

Put the learners into small groups, and tell them they are going to make a holiday destination guide to their town or city:
- *What do they think should go in the video?*

Once they have brainstormed their list of sites and attractions, ask them to 'storyboard' the video, putting the sites in order, adding spoken explanations to each of them, choosing music, and so forth.
- If it is possible, get the learners to collect real video materials – by walking around the town/city and filming them.
- If it is not possible, ask them to collect a set of images or video clips from the web – to illustrate their video.

The learners put the videos together:
- They do this, using a video production app.
- They combine photos, any video clips, title screens, music and (where possible) audio commentary into the finished product.

Watch the videos together and vote on the best.

If you have a class blog or YouTube channel, upload the video and share it.

Run on

Ask the learners to create a *second* tourist video for a place in the world they would *like* to visit.

Hold a holiday 'showcase event' in class, and rank the places in order of preference for a holiday.

Acknowledgement: Our thanks to Anglo European School of English in Bournemouth, UK, for allowing us to use their student-created movies as examples for this activity.

This is my song

Music videos

The learners share their favourite videos.

Run up

Find a music video you like, and prepare a short explanation about it.

Run

Start by talking a little about your music video likes and dislikes – try to address all the talking points in the speaking activity below.

Show the learners your current favourite music video – try to give them as much information about the singer/group, the song and the video as possible.

When you have finished showing the video, invite questions.

Move on to a general discussion activity about music and music likes and dislikes. The learners do this in pairs, along the following lines (tailoring the questions to your class):

- *What sort of music videos do they like?*
- *What sort of music videos do they really not like?*
- *Who are their favourite video music artists?*
- *Which of the groups they like make good videos?*
- *What was the last music video they saw?*
- *What are their favourite music videos of all time?*
- *What's their current favourite music video?*

Now tell them they are going to share their own current favourite music video with their partner.

Ask them to find the video on YouTube or the artist's official website, or similar:

- In pairs, the learners show each other their video and explain a little bit about it – the artist, the song, the video, and anything else they know about the song.
- Their partners make notes on what they are told, using a note taking app on their mobile device.

Encourage the pairs to ask questions, to get as much information for their notes as possible.

When they have finished showing each other their music videos, get them to mail the notes they took to their partner. They will need them for the 'Run on' activity.

Run on

Ask the learners to check their mail when they get home, and to look at the notes their partner sent them.

They record a piece to camera, on their mobile device, about their favourite music video, remembering to include the following information:

- The artist, album and song title (where appropriate).
- A description of the video itself.
- Why they like it.
- How to find the video online.

Collect all the videos in the next lesson and add them to a class YouTube channel, blog or wiki.

Invite the learners to browse the videos and leave comments and questions.

Show and tell

Pecha Kucha presentations

The learners create a short presentation
on a hobby or interest.

Run up

Film a Pecha Kucha presentation around a hobby or interest
of yours. You can find out about some 'frequently asked
questions' here:

http://www.pechakucha.org/faq

If you can't film yourself doing the Pecha Kucha, you can do
your presentation live in class.

Ensure that the learners have a video image app on their
devices.

Run

Write the subject of your Pecha Kucha on the board, and
tell the learners they are going to watch a presentation
connected to this theme.

Put them into small groups, to brainstorm:
- *What information might they expect to hear in the
  presentation?*

Show them your presentation (or perform it for them live)
and ask them to note:
- Things they expected to hear about, which were there.
- Things they expected to hear about, which weren't there.
- Any questions they have about the subject.

After the presentation, conduct feedback as a whole class
and answer any questions they may have.

Tell the learners a little about the history of Pecha Kucha
(see opposite) – the structure and format.

Tell them they are each going to prepare a Pecha Kucha on
a subject of their choice:
- They decide on the subject.
- They sketch out a storyboard for their presentation.
- They can do the storyboard on paper, or using a note
  taking or drawing app on their mobile device.

Put the learners into pairs, to run through their storyboards
– helping each other to refine the structure and content.

They record their Pecha Kuchas, using a video image app:
- They upload them to YouTube or another file sharing
  website.
- They can also share their videos on a class blog or wiki,
  if you have them.

Ask each learner to view two of the others' Pecha Kucha
presentations, and leave some comments, feedback or
questions.

Pecha Kucha

A Pecha Kucha is a presentation in which there are
twenty slides:

- Each slide is shown for twenty seconds.
- The speaker has no control over the slides,
  as they are set to advance every twenty seconds.
- Pecha Kuchas force speakers to be organised and
  succinct, and remain calm and collected.

The Pecha Kucha was invented by two architects
in Japan:

- The first 'Pecha Kucha night' was in Tokyo in 2003.
- Pecha Kucha nights are now held in over 700 cities
  around the world.
- They are held in in bars, restaurants, clubs, homes,
  studios …

Run on

Ask the learners to choose a Pecha Kucha from the official
Pecha Kucha website:

http://www.pechakucha.org/watch

They should find one that is in an area that interests them
personally:
- *Why did they choose it?*
- *What did they like about it?*
- *What would they have added if they had made it?*

They share it with the class, either in class or on the class
blog or wiki.

Trail blazer

Film trailers

The learners create a trailer for an imaginary film.

Run up

You may want to give your learners an example of a student-created film trailer. For an example of what to aim for, see: *http://goo.gl/mD1brh*

Note that this was made with iMovie on an iPad, an app that does much of the work in making the end result look professional.

Run

Talk about a film you saw recently – plot, actors, main story – and what you liked and disliked about it.

Ask the learners if they saw the trailer for the film you talked about:
- *If they did – what information was in the trailer?*
- *If they didn't – can they imagine what the trailer was like?*

The learners brainstorm the main features of a good trailer:
- Suitable music
- Titles and subtitles
- Actors
- Overview of the story
- Two or three short scenes to grab audience interest
- Basic information (where and when showing, etc)

Play your trailer to the learners:
- *Did it feature all the aspects in the list above?*
- *Was anything missing?*
- *What did they think of the trailer?*

Tell the learners they are going to make a trailer for a new film, starring *them*.
- They should decide what genre of film they would like to be in.
- They should sketch out the rough story, including:
  The title and credits
  The music
  The scenes they are going to film
- They should decide *who* will be responsible for collecting *what*.

When they are ready, they can proceed to the 'Run on' stage of the activity.

Run on

Ask each group to go and film the scenes they need, and to collect other resources such as photographs and music if needed.

This should be done before the next lesson, and saved to their mobile devices.

In the next lesson, they will work together to make the finished trailer, using a video production app.
- For iMovie on iOS, they can go right ahead and film the trailer using iMovie, as this is one of the built-in templates.
- For Android, options are a little more limited, so it may be better to transfer the 'assets' (video files, music, etc) to desktop or laptop computers and do the editing there, using iMovie or Windows Movie Maker.

Organise a 'showcase' of the finished trailers, and vote on the film the class would most like to see.

Acknowledgement: Our thanks to Anglo European School of English in Bournemouth, UK, for allowing us to use their student-created movie as an example for this activity.

Book bonanza

Video book trailers

The learners create a video trailer of a favourite book.

Run up

Find book trailers online, to show to your learners. For example, at:
http://goo.gl/jb3UH

You may also want to create your own book trailer and prepare a handout of guidelines.

Run

Write *Book trailer* on the board and ask the learners if they know what it is. (A book trailer is a video preview or summary of a printed book.)

If necessary, give the learners a clue by asking them to think of a 'movie trailer'.

Ask the learners to brainstorm what content a typical book trailer video might include:

- Background music
- Read-aloud extracts
- Images
- Key words
- An interview with the author …

Put their suggestions on the board.

Show two or three professional book trailers of different styles. See the 'Run up' above.

Reassure the learners that they don't need to understand everything – they should look at the general style and content of these videos, as a film-maker would.

Refer to the list of brainstormed elements on the board:

- *What elements are in these book trailers?*
- *What other elements are included?*
- *Which trailers do they like best? Why?*
- *Which are the most difficult to make?*
- *Which are the easiest to make?*

Tell the learners that they don't need to be a professional film-maker to produce a good book trailer.

Show two or three trailers produced by third-grade and fourth-grade learners – ie between the ages of 8 and 10 – (from *http://goo.gl/jk8AF*). These are all produced with a simple video production app (Animoto).

- *Can they guess what each book is about from the trailer?*

Put the learners into groups of three to produce a book trailer, and give them steps (on the board or in a handout) to help them. See opposite.

Making a book trailer

Step 1

Choose a book that you have all read and like.

Step 2

Find images and background music for your book trailer.

Step 3

Choose some key words and phrases that summarise the book.

Step 4

Audio- or video-interview one of your group as 'the author', or create an audio voiceover about the story. (This step is optional.)

Step 5

Create a storyboard with all of these elements – remember to include the book title at the beginning!

Step 6

Now create your video on a mobile device with a video production app.

If they use a web-based video production app (such as Animoto), it is easy to access and share the book trailers online.

Watch the finished trailers as a class. Ask the learners:

- *Have they read any of these books?*
- *If so, do the trailers reflect the content?*
- *For new books, which ones do they most want to read, based on the trailers?*

Run on

For homework, ask the learners to upload or link to their book trailers in a class blog.

Encourage them to watch each other's trailers again, and to leave a text comment on each trailer.

Going Mobile has so far examined how you and your learners can get started with using mobile and hand-held devices. We outlined some of the options available (Part A), and suggested a number of activities to carry out using mobile devices, both inside and outside the classroom (Part B).

Part C aims to demonstrate how you can 'go further' in the context of your classroom, and also 'go further' in the context of your institution. It consists of two sections:
- First we look at how to experiment with more prolonged activities and projects – as you become familiar with a fuller range of mobile possibilities.
- Then we look at how to develop a step-by-step implementation plan for the use of mobile devices – at an institutional level.

Going further ... in your classes

These activities build on those in Part B, and combine many of the elements we examined there (text, image, audio, video) into longer and more demanding tasks for learners.
- The activities encourage your learners not only to use their devices both inside and outside the classroom (as was often the case in Part B), but also to exploit the special features that smart devices offer, such as geolocation and augmented reality.
- They encourage your learners to work with varied and rich *input* from their surroundings, as well as to produce equally varied and rich *output*, both in terms of language and media.

Many of these activities will take several hours of class time to implement, so they are best carried out as short projects over several lessons. We hope that, if you and your learners have already tried several of the activities in each of the chapters in Part B, you will already be in a position to approach these *'Going further ...'* activities with confidence.

Going further ... in your institution

Of course, you could easily try out any of the activities in this book with your own learners 'now and again', working independently.
- But it makes more sense to take a broader view – to integrate the use of mobile devices into a clearly defined pedagogical framework.
- And it makes much more sense to work as part of a team – to explore how mobile devices can be most effectively used by many (or all) of the teachers in your institution.

This will provide *you* with more support, and also provide a more coherent approach and learning experience for *your learners*.

This wider approach – of developing beyond the confines of the individual teacher or the individual class, and learning exactly how to carry this development out – is what forms the pedagogical philosophy of Part C.

Going further ... in your classes

It should be clear by now that mobile devices have a special quality – they are, well, *mobile*.

And as such, they have the power to take learning experiences beyond the classroom walls.
- In Part A, we pointed out that *devices* can be mobile, *learners* could be mobile, and *learning experiences* could be mobile.
- In Part B, we described activities that often include a 'going mobile' stage, with learners working with devices while moving around inside and/or outside of the classroom.
- In Part C, we focus particularly on *learning experiences* which are mobile, and require learners to do a significant amount of work beyond the classroom. We also focus on learning experiences that use the special 'affordances' (or properties) of smart mobile devices, such as tactile screens, geolocation and augmented reality.

The activities in this section encourage the learners to create multimedia with their devices, producing text, images, audio and/or video. In this way, these activities build on the activities described in Part B.

As well as requiring learners to do tasks with mobile devices mainly outside the classroom, several activities need to take place over a number of lessons:
- You can set up an activity in one lesson.
- The learners then carry out a part of the activity outside the classroom.
- In the next lesson, you can get them to work further with the information they have collected, using appropriate language.

A few of the activities are designed to practise a specific grammar point or specific vocabulary. For eaxample:
- *Living history* practises present tenses.
- *Tap that word* reviews vocabulary sets.

However, most of the activities require a more task-based approach – the language needed by the learners will *emerge* as they work through the activity.

That's not to say we can't *predict* the sorts of structures that the learners may need. But rather than pre-teaching these structures, and insisting on the learners incorporating them while doing the activities, we provide them with the necessary language at the point of need.

Be prepared!

Before we look at these activities, there are a few things you should bear in mind at the preparation stage, *before* you start any mobile-based out-of-class work with your learners. They concern these key areas:

Places

Think about exactly *where* your learners will physically need to go:
- *Will they stay within the school building or grounds?*
- *Will they need to move around your town or city?*

If you are working with learners under 18, you may need to get written permission from their parents to take them off the school grounds, so check with your school for rules relating to this.

Consider, also, exactly *when* your learners are going to be moving around to these places:
- *Will you ask your learners to carry out their mobile project work in their free time between classes, or over a weekend?*
- *Will you include the work as part of a school trip, or as class time spent outside of the school building?*

The activities in this section are suitable for learners aged about 13 and upwards.

We also suggest variations for young learners at times. However, there is no reason why you can't carry out some of them with young learners as part of a class trip, providing you have the necessary permissions to take them off the school premises.

People

Consider who the learners will need to talk to or interview as part of the activity.

You may want to get permission from these people in advance, especially if your learners need to take photos, or audio- or video-record their interviews.

Although filming in a crowded public place – such as a park or station – may be acceptable, filming someone in an interview is a different matter. You and your learners need to identify:

- *What other people will be directly involved in the project?*
- *What media recordings will be taken of these people (photos, audio or video)?*
- *Who has copyright over the material?*
- *What level of privacy will be applied to sharing the material?*
- *What steps will be taken to protect the subjects' data, such as protecting their identities?*

You may find it useful to produce a simple 'permissions form' for interview subjects to sign, once their rights – and the data protection measures you and the learners will take – have been clearly explained.

Devices

The activities described in this section require smart devices. You will need to decide:

- *Will the learners use one smart device between two (or in a small group)?*
- *Will they work individually with one device each?*
- *Will they use their own devices?*
- *Will they use class sets supplied by the school?*
  (See Part A for more on this.)

If your learners are going to use class set devices, remember to check that the school's insurance policy covers possible breakage, loss or theft when the devices are taken off the school premises!

Some of the activities also require internet connections out of class, although in some, the learners can use the school wifi connection to complete the activity if they are unable to use the internet while on the move.

Dissemination

If your learners are going to invest time and effort in creating multimedia with their mobile devices, it makes sense to share what they produce with others.

So, before you start, think about:

- *Who will your learners share their finished work with?*
- *Will it be shared with their own classmates, other classes in the school, parents (for younger learners), other schools, the local community – or even with the world?*

Depending on what community their work will be shared with:

- *It might best be done face-to-face – for example, during a school project day.*
- *It might best be done online – for example, via a class blog.*

And if the work is going to be shared online, you need to consider what levels of privacy and access will be applied.

For example, if you and the learners decide to share their work via a class or school blog:

- *Will the blog be open to anyone to visit and to leave comments on?*
- *Will access to the blog be restricted, and if so, to whom?*

And remember: if any images of people or interviews are shared online, the necessary permissions need to be sought first (see 'People' above).

Assessment

Finally, you must decide whether you are going to assess your learners' work. For example, the project may form part of an end-of-term assessment, instead of an exam:

- You will need to let your learners know they are going to be assessed.
- You will also need to be very clear on the criteria you are going to use for assessment, and share these criteria with your learners in advance.

As we mentioned in Part A, if your learners are producing multimedia, then you may want to include assessment criteria that take this into account, and not just focus on the language produced (whether written or spoken):

- You may decide to give the learners credit for how they present their activities and how engaging or effective the final multimedia format is.
- You may decide to comment on their choice of visuals, their use of hyperlinking, or whether they correctly source any extra images or background music.

You may also decide to give the learners points for the *process* aspects of their project, not just the final product. 'Process' aspects include:

- *How well did the collaboration between group members work (for a group project)?*
- *How much planning and preparation went into the project?*
- *How well were the project content and aims communicated to any third parties involved?*

As a rule of thumb, we would suggest:

- It's a good idea to first integrate the assessment of shorter mobile-based activities (such as those described in Part B) into your overall assessment of your learners, before you assess more complex digital work.
- It's not a good idea to suddenly launch into assessment of a larger-scale digital project, if your learners have had no previous experience of having their digital work assessed.

Go further!

We now look at 10 mobile-based activities. We have divided these into four different types:

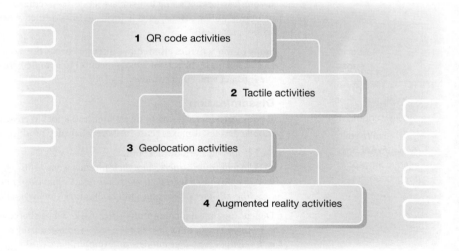

1 QR code activities

2 Tactile activities

3 Geolocation activities

4 Augmented reality activities

As in Part B:

- We have ordered these activities in increasing order of complexity, using the SAMR model as a guiding principle (see Part A, page 26).
- We move from more 'text'-based activities to those using more complex mobile device functions.
- The QR code and tactile activities correspond to the 'enhancement' stage of the SAMR model, while the geolocation and augmented reality activities correspond to the 'transformation' stage.

QR codes

These two activities provide a staged introduction to the 'reading' of QR (Quick Response) codes. QR codes are similar in concept to supermarket barcodes: information is encoded in a two-dimensional graphic:

- A barcode consists of vertical lines.
- A QR code encodes information vertically and horizontally – so you can get more information on it.

You read a QR code by scanning it with a QR code reader app:

- The app uses the camera on your mobile device.
- Typically, a code will include a short amount of text (up to a couple of lines), or simply a website link (URL).
- QR codes can also link to an audio or video file.

If the device is internet-enabled, you can immediately click on the URL and go to the web page or audio/video file.

Coded review

Scavenger hunt

Coded review
An introduction to QR codes

The learners use a QR code reader to read and answer review questions.

The activity takes place within the classroom.

Get the app

- You will need to use an online QR code *generator* to create QR codes for your learners to read. We recommend Kaywa or The QR Code Generator:
  http://qrcode.kaywa.com/
  https://www.the-qrcode-generator.com/
- You and your learners will also need a QR code *reader*. We recommend i-nigma or Scan (see page 30).

Get ready

Prepare four to six questions that review recent language covered in class. For example: with a group of beginners or elementary learners, you could use short focused questions like these:

- *What are your three favourite sports?*
- *What are two of your favourite foods?*
- *What do you enjoy doing on weekends?*
- *What kind of music do you like?*
- *What kind of films do you like?*
- *What do you like doing in the English class?*

Create a QR code for each question:

- Go to your chosen QR code generator web page – see 'Get the app' above.
- Choose the QR code format ('text' in this case).
- Type in a question.

Click on 'Create', and then print the QR code. Do this for all your questions.

Take several strips of paper to class for the learners to write the answers to your questions.

Get going

Show the class one of the QR codes and ask:

- *Have you seen these codes before?*
- *Where have you seen them?*

The learners open their QR code reader apps on their devices, and read the QR code you are showing them.

Put all your QR codes around the classroom walls, and put the strips of paper in the centre of the room.

Ask the learners to work individually, and to walk round the classroom, reading each QR code message with their QR code readers. For each question:

- They pick up a strip of paper and write their answer to that question.
- They stick their answer on the wall, next to the QR code.

Tell them *not* to write their *names* with their answers!

Once most (or all) of the answers are on the walls, the learners walk round again and read the answers.

Nominate learners to each take down one QR code and its surrounding strips of paper.

- They read out the QR questions and the responses.
- They pause after each response, so that the class can guess *who* has answered *what*.

In pairs, the learners produce spoken or written sentences about their classmates from what they can remember. For example:

- *Saud enjoys going to the cinema on weekends.*
- *Ilya's favourite foods are Italian and Chinese food.*

Going mobile

When you go mobile by putting QR codes on the classroom walls, a structured review activity like this one is more fun and motivating. There are photos on Nicky's blog of a group of learners carrying out this activity:
http://goo.gl/HRUr5B

You can also use QR codes to link to specific URLs – the learners are then taken to websites, and you can base your tasks around these. For example:

- You choose four websites related to your current coursebook topic.
- The learners read each QR code.
- They visit each site.
- They note down two things they have learned about the topic from the site.

This activity introduces a useful digital skill – that of knowing how to read QR codes – for learners *and* teachers!

▌▌▌▌▌▌▌

Scavenger hunt
Getting to know the school

The learners read QR code instructions and carry out a number of tasks, covering all four language skills and using a range of mobile device functions (reading QR codes, taking photos, audio recording).

The activity takes place within the school, but the 'Going mobile' suggestion gets the learners reading QR codes out of the classroom.

Get the app

- You will need to use an online QR code *generator* to create QR codes for your learners to read. We recommend Kaywa

or The QR Code Generator:
http://qrcode.kaywa.com/
https://www.the-qrcode-generator.com/

- You and your learners will also need a QR code *reader*. We recommend i-nigma or Scan (see page 30).

Get ready

Prepare 10 questions, based on various locations around your school.

- Each question should relate to a specific location in the school.
- Some should require the learners to speak to people in that location, to take a photo, to note things down, or to audio-record.
- Each question should also tell the learners where to go next.

On page 99 are the sample questions we used with an elementary class, based on locations in the school where we were working. Question 10 is not really a question, but 'closes' the activity.

Before you create your questions:

- Check that any people potentially involved (eg for interviews) are happy to take part or to be photographed.
- Check that they will be available during your class time, to answer your learners' questions.

Generate, and print out, 10 QR codes – one code containing each question.

- Note the question number on the code (or write the question in full on the back) so you can easily identity it.
- Before class, go to each location and stick the relevant QR code question on the wall.

Get going

Give the learners a grid with each location for the 'scavenger hunt' on a handout (or draw it on the board). Our example grid looked like this:

Scavenger hunt answers

Question	Where?	Answer
1	Our class	
2	Social events office	
3	Dining room	
4	Dining room	
5	Lecture theatre	
6	Reception	
7	Library	
8	Room 11	
9	Our classroom	
10	Our classroom	

Show the learners the QR code for Question 1 and ensure that they can read it with the QR code reader on their devices.

- Ask them to note down their answers on their grids.
- Point out that the QR code also tells them where to go next (it is also on the grid).

Send the pairs around the school to read the QR code question on the wall in each place.

- They complete each task.
- Depending on the size of the class and the complexity of the tasks, you can give them a time limit (eg 30 minutes) after which they need to come back to the classroom – whether they have finished or not.

While they are working around the school, you may want to visit each location yourself – to ensure that they are on track, and to help out as necessary.

When the learners are all back in the classroom, regroup them to compare their answers, then check the answers with the whole class.

For homework, they listen to any audio recordings they had made (see Questions 8 and 9 opposite) and write a short paragraph about each of the people interviewed.

Going mobile

Going mobile with QR codes around the school encourages the learners to explore and get to know the school in a fun and challenging way. Nicky's blog contains photos of a group of learners carrying out this activity in a school in the UK: *http://goo.gl/HRUr5B*

If the learners have access to internet-enabled devices out of class (and you live in a small town!) you could put QR codes in public places, with questions designed around each location. For example, you could put QR codes in the train station with questions like:

- *What time is the first train to x?*
- *How much is a ticket to y?*

The learners can also be encouraged to each produce a single QR code with a question or task themselves, in which case they will also need a QR code generator – see 'Get the app'.

- They send it to you by email.
- You print them out, and put them on the classroom walls for the next lesson.
- The learners then carry out their peers' QR tasks.

This activity develops the learners' digital skills with QR codes by setting them more challenging multimedia tasks (such as recording audio), and finally by asking them to produce their own QR codes for their classmates to read.

Scavenger hunt questions

Code 1
What three things do you enjoy doing on weekends? Write them down.
Next: Social events office

Code 2
Social events office:
What activities are on in our city this weekend? Write down two you like.
Next: Dining room

Code 3
Dining room:
What is for lunch today? How much is it? Take a photo of the menu.
Next: Dining room

Code 4
Dining room:
Take photos of three other foods you see.
Next: Lecture theatre

Code 5
Lecture theatre:
How many chairs are there in the room? What colour are they?
Next: Reception

Code 6
Reception:
Ask two people in Reception for their full names, and write them down. Ask to take a photo of them. Ask for permission first – politely!
Next: Library

Code 7
Library:
Find the elementary graded readers. Write down three book titles.
Next: Go to Room 11.

Code 8
Room 11:
Jane is the Director of Studies. Ask Jane three questions: about her family, her job, and what she enjoys doing. Prepare your questions first, and practise. Audio-record the interview.
Next: Our classroom

Code 9
Our classroom:
Ask Nicky three questions: about her family, her job, and what she enjoys doing. Audio-record the interview.
Next: Go to Question 10.

Code 10
You're finished! Well done!

Tactile screens

Smartphones have tactile screens – that is, they respond to touch. Many mobile games and apps rely on this functionality.

- The two activities described here both rely specifically on touch to work.
- They also integrate text, audio and video, to create rich learner-created multimedia experiences.

Learners of all ages respond well to activities that revolve around the unique features (the unique affordances) of tactile screens:

- Adults enjoy the game-like quality of language activities that rely on the touch screen.
- Young learners who are still developing their reading skills should find tactile activities that use audio and images or very simple individual words especially effective.

If the learners have access to internet-enabled devices, they can create and share tactile games – not just with each other, but with the wider community.

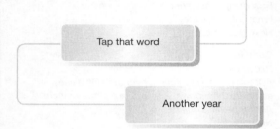

Tap that word

Another year

Tap that word
Tactile vocabulary

There are apps that enable learners to create games for mobile devices and that rely on the tactile interface of smart devices. This activity encourages the learners to make simple games that review vocabulary areas, and to play the games outside class.

Get the app

The app in this activity – Tiny Tap – is aimed at children, but adults too can create activities and make games for other learners.

- The games use a background picture or photo, and the learners overlay audio clues and drawings.
- The players listen to a clue, then tap the part of the screen referred to in the clue.

You can currently download Tiny Tap free from the iOS App Store (at the time of writing, an Android version is in development). There are also several other tactile game apps that can be used to create quizzes, and have similar functionality (see page 30).

Get ready

Once you have downloaded the app, try one or two of the free games offered with it, to see how it works. Then create your own simple game by clicking on 'Create' and following the steps provided by the app.

Below is an example, based on 'Countries, nationalities and languages':

- Use an image of a world map (with only the names of countries written on it) as your background image for the game.
- Add audio clues about different countries:
  Where do people speak Dutch?
  What country's capital is Moscow?
  Where can you see the Eiffel Tower?
  What's the largest country in South America?
  What country in Europe has four official languages?

Add your game to the Tiny Tap 'marketplace', and make it free.

Get going

The learners download the app and play the countries/nationalities game that you created.

Now ask them to create their own games in pairs – to review a vocabulary area – and give it a name:

- They will first need to take, or find, a background picture.
- They will then place five or six audio clues on it.

They can review an area of vocabulary you have covered in class, or here are some common themes to choose from:

- Stationery: *pen, ruler, notebook, sharpener, rubber, stapler …*
- Body: *arm, leg, stomach, ankle, foot, hand, knee …*
- Face: *eye, nose, mouth, ear, hair, eyebrow, cheek …*

- Landscape: *house, tree, hill, fence, grass, flower …*
- House: *bedroom, kitchen, hall, dining room …*
- Furniture: *chair. sofa, lamp, table, carpet …*
- Clothes: *skirt, shirt, socks, hat, scarf …*

Give the learners time to prepare and rehearse their audio clues – before they record them in the app on their mobile devices.

Once they are happy with the final version of their game, they save it and upload it to the Tiny Tap marketplace.

Share the names of all the games with the class – so they can find them easily in the marketplace – and ask them to download their classmates' games to their devices.

Give the learners a few days to try out all the games.

They can play these games at any time or place that suits them best – for example: while commuting to school or work, while doing errands or waiting for something, at home …

Back in class, get feedback:
- *Which games did you enjoy the most?*
- *Which games were the easiest, and the most difficult?*
- *Where did you play the games?*
- *How often did you play them?*

Ask the learners to keep an eye on the Tiny Tap marketplace – to see how many people download and play their game.

Check this with them a few weeks later, pointing out that they have created a game that could in theory be played by many people all over the world!

Going mobile

When you go mobile by creating a tactile game (in this case, games or quizzes that review vocabulary) you are encouraging the learners to create and review learning materials in a way that many find very motivating. Tiny Tap allows the learners to create audio clues to review language, but other apps (see page 30) enable learners to create text clues, or to match text to images.

The age and level of your learners will determine the language they use in their clues:
- If the learners are very young or have a low level of language proficiency, the clues they create can be very simple (*Tap the pen! Find the ruler!*).
- If they are more proficient, they can create more complex clues (*What do we use to write with? What helps us draw straight lines on a page?*).

This activity develops learners' gaming literacy in an educational context, but not just by asking them to consume games created by others. It also helps them develop the skills to create their *own* multimedia tactile games and quizzes, while reviewing language at the same time.

▌▌▌▌▌▌▌

Another year
A multimedia yearbook

In this extended activity, the learners create a touchable yearbook featuring a variety of media which they collect from their classmates. The end result is a 'class photograph' which is both interactive and multimedia-rich.

Get the app

The learners need to install ThingLink:
http://www.thinglink.com

ThingLink is a web service that allows you to upload photographs and put 'hot spots' on them. Once clicked, these hot spots can launch a variety of media: text, websites, video, audio – and more.

There are also a number of other apps that enable learners to record audio with photos (see pages 29–30).

Get ready

The web version of the service is complete, whilst the app (currently only available for iOS) has a reduced feature set, best used for viewing rather than creating.
- Register an account at the ThingLink website, and upload a class photograph for the learners to tag.
- Ensure, when explaining the project to the learners, that you have a ready-made example to show them, so try to get one made in advance, with the major affordances exemplified (web link, audio, video, photograph …).
- You can watch this tutorial if you need help with creating your example:
http://goo.gl/2iOzdF
- Ensure the learners will have access during class to computers rather than mobile devices – to upload their content and add it to the class photo on ThingLink.

Get going

Divide the learners into pairs.

Tell them that they are going to make a multimedia yearbook for the class:
- Show them the photo you uploaded to ThingLink.
- Explain how ThingLink works.
- Show them your example.

Tell each pair that they are responsible for the content that will be associated with their partner. This content could be:
- A video interview.
- An audio interview (or partner's favourite sounds, etc).
- Links to favourite websites.
- Links to social media.
- Photographs.
- Google Maps.
- A Wikipedia article.

For a complete list of possible content that can be added to ThingLink, see:
http://goo.gl/gcswNP

The pairs plan what each of them will share with their partner:

- They script and rehearse an audio or video interview.
- They identify two or three photos to share.
- They plan their overall profile to share with their partner. For example:
  Their favourite websites.
  Their social media profiles.
  A favourite place.
  A Wikipedia article on a topic they are really interested in.

Tell them to collect at least five pieces of information about their partner in total. This should include:

- An audio or video interview (filmed on their devices).
- Photos (taken on their devices if necessary).
- Three other pieces of information from the list above.

Ask each learner to add the media and information for their partner to the class photo in ThingLink, using computers (not the mobile app).

Once the class photo, is complete, the learners explore the hot spots on the class photo to find out more about all of their classmates.

Going mobile

This activity goes mobile by using the touch screen affordances of smart devices to allow the learners to create and share information (about themselves in this case) in a rich multimedia presentation format.

A variation on this activity has the learners collecting multimedia outside class, and then creating their own individual ThingLink pages to share:

- They can collect images, audio and/or video from one important historical place in their town or city, and add it to ThingLink along with related Wikipedia articles.
- They can collect images, audio and/or video from their top three favourite restaurants in their town or city, and add this to ThingLink along with the restaurants' locations on Google Maps, and social media links such as the restaurants' Facebook pages and Twitter feeds.

Either of these ThingLinks can easily be shared with the wider community, as they contain public rather than personal information.

This activity develops the learners' multimedia literacy by requiring them to produce images, audio and/or video for a specific purpose.

The class yearbook activity also encourages the learners to think about the image they are projecting of themselves to their peers, and can result in a useful conversation about online identity management.

Geolocation

Geolocation tells us where an object or person is, using digital technology such as a GPS signal or an internet signal. These three activities encourage the learners to use their GPS-enabled devices in specific geographical locations. For example:

- They may need to visit a specific location, to find out something about it using their mobile devices.
- They may need to create text or audio information at a specific geographical point, and share that with others.

The activities in this section all rely on learners using GPS-enabled smart devices to carry out tasks:

- The first two activities encourage the learners to create their own audio recordings and to pin these to an online map: this is 'geotagging'.
- The third activity involves hiding physical objects in specific geographical locations and then sharing the GPS coordinates.

The learners create digital geolocated objects (such as audio recordings), and they create – and hide – real physical geolocated objects.

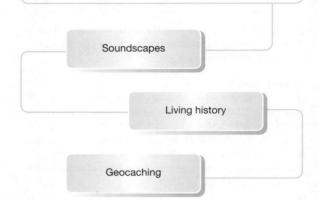

Soundscapes

Living history

Geocaching

Soundscapes
Mapping the sounds of the city

Sound is an evocative medium, and it is very easy to capture with mobile devices. This activity gets the learners to create a soundscape of their town/city.

They visit places that are significant to them, record ambient sound, and add these audio recordings – with written comments – to an online map.

Get the app
The learners will need to install Woices (*http://woices.com*) on their mobile devices and register for an account, before starting this activity. Woices allows users:
- To create short audio recordings (called 'echoes').
- To add text and key words to their recordings.
- To save it to a specific geographical location on a map.

Get ready
Put the learners into pairs or small groups and ask them to discuss the following questions:
- *What are your favourite places in this town/city?*
- *Why do you like them?*
- *When you visit them, what sounds do you typically hear?*

Ask each group to briefly share their responses with the class. You can also share *your* favourite places:
- Special cafe or pub
- Park
- Busy shopping street
- Famous monument
- Shop
- Museum
- Port or river …

Tell the learners they are going to each create an individualised soundscape of the town/city.

Get going
Ask the learners to visit four of their favourite places, and to record 30 seconds of ambient sound in each place on their mobile devices. They should not speak while recording.
- If they can connect to the internet via their devices while on the move, they can immediately add each of their 30-second recordings to its geographical location on an online map in Woices.
- If they can't connect to the internet, they should save their recordings on their devices – and then add them to Woices using wifi, either when next in the classroom or at home.

Woices also allows for typed text to be included along with each audio recording, and the learners should add at least 50 words for each recording, explaining:
- *Why is this place significant to them?*
- *Why do they like it?*
- *What things can be heard in the recording?*

As you will be sharing the recordings with the learners in a subsequent lesson, ask them to tag (label) each of their recordings with the same keyword (eg 'Class5') – it will be easy to search Woices for these recordings later, and to listen to them with the class.

In the next lesson, listen to the recordings with the class:
- You provide individualised feedback on the written texts accompanying the learners' recordings.
- You provide feedback on common errors with the whole class.

The learners can then edit their texts in Woices, to improve them.

Going mobile
Going mobile by asking your learners to record ambient sounds outside the classroom, in places that are important to them, encourages them to pay attention to their surroundings. It also elicits an emotional response to these places, which can then be reflected in the rich accompanying texts that they produce.

You can ask the learners to create soundscapes around specific themes. For example:
- City soundscapes based on a special day: Halloween, St Valentine's Day, Independence Day …
- Soundscapes based on a specific theme: food, traffic, sport, going out …

If you work with young learners, they can create soundscapes of the school by recording ambient sound in the cafeteria, library, reception area, playground …

This activity provides an original way for the learners to bring personalised audio content from the outside world into the classroom, which then functions as a prompt for more focused language work (in the accompanying texts).

Living history
Making the past present

'First person' accounts of historical events can make the past feel much closer. The learners research an important historical event that took place in their town or city, and then imagine that they were eyewitnesses to this event.

They audio-record their impressions of the event, take photos of the place, and then upload the photos and their recordings to an online map.

Get the app
The learners will need to install Woices (*http://woices.com*) on their mobile devices and register for an account, before starting this activity.

If your learners have already done the previous activity to this one (*Soundscapes*) they will be familiar with Woices and what it can do.

Get ready

Research a very well-known international historical event, and create an eyewitness audio recording in Woices (see 'Get going' below, to see what information to include). You will play this recording to the learners in class.

Get going

Explain that you are going to revisit history:
- Mention the historical event you chose and ask the learners what they know about it.
- Play your Woices recording, so they can check if they are correct.
- Explain that this is a model for what they will produce.

Ask the learners:
- *What important historical events took place in our town?*

In pairs, ask them to use the internet to research two or three important local events. For each one, the learners should make notes:
- *When and where did the event take place?*
- *Who was involved?*
- *What was it about?*
- *Why was it important?*

Ask the learners to tell the class about their findings, and add a list of the events to the board.

The class should agree on *one* event and research it in detail.

They then imagine they are *witnessing* the event:
- They prepare an eyewitness account of the event as it unfolds before their eyes.
- They use present tenses to describe the scene:
  It's a cold rainy day …
  The crowd is running towards me …

You could put a few prompts on the board to help them:
- *What is the weather like?*
- *What can you see, hear and smell?*
- *Who are the people around you? What are they wearing? What is their mood?*
- *What happens? Describe the event.*
- *What happens next?*
- *How do you feel?*

Help with language, and give the learners plenty of time to make their accounts detailed and engaging – as if they really are there!

Ask the learners to rehearse their accounts, and when they are ready, to find a quiet place to record in Woices, and to save their accounts to the geographical location where the event took place.

The learners now visit the place where the event unfolded:
- They take a few photos of the place.
- They add their photos to their own recordings in Woices.

Tell them to search for the audio recordings created by their classmates as eyewitnesses – and to listen to these while they

are still standing at the same place.

Back in the classroom, ask:
- *Do the eyewitness accounts show the historical event from different perspectives?*
- *To what extent has it made the past feel more present to you?*

Going mobile

The learners literally 'go mobile' in this activity when they take photos at the site of the historical event, as the final stage. Their eyewitness audio recordings are first prepared and rehearsed (which is when *you* need to be on hand to help with language and pronunciation). They are then recorded, and uploaded to Woices.

Rather than getting all the learners to create eyewitness accounts of the same historical event, here are some alternatives:
- The learners work in pairs, and choose a different historical event and place in the town/city. They prepare and record their eyewitness account together, and listen to their classmates' eyewitness recordings in a subsequent lesson, or for homework.
- The learners work in pairs or individually, and choose an international historical event for their accounts. This version can be done entirely in the classroom.

This activity helps the learners understand geotagging by asking them to add recordings and photos to specific map locations. Understanding geotagging is a basic digital literacy.

▌▌▌▌▌▌▌

Geocaching
Treasure boxes

'Geocaching' is an activity that gets participants to find objects (or 'caches') which have been hidden in the real world, by using GPS coordinates. A typical geocache is a waterproof box containing a logbook and information or small objects.

In this activity, the learners create a geocache box, hide it, and log their caches on a geocaching community site.

Get the app

There are several geocaching apps, and we recommend OpenCaching (see page 29).

OpenCaching supports a large network of geocachers:
- The users can try out others' geocaches and also upload their own, at no charge.
- The app comes with a thorough initial explanation of geocaching.

Get ready

Ask the learners to download the OpenCaching app to their mobile devices and read the introductory screens, which explain how geocaching works (or you explain this to them).

Get going

Ask the learners to open the map icon in the app:

- *Are there any geocaches in the area?*
- *If not, are there caches on the map in regions nearby?*

Give them some time to look at several examples of caches on the map, to get an idea of how they work:

- They can click on a map point, read the description, and also reveal hints for some caches.
- They can click on 'Log activity' and read comments from users who found (or didn't find!) the cache.

In groups of three or four, the learners plan what objects to put in their geocache.

- All physical geocaches should have a logbook (notebook) and pen or pencil, so that finders can leave comments with the cache, for the next finder to read.
- The learners can search online for advice on what to put in a geocache – geocachers often put in inexpensive items, or second-hand books or CDs, for others to find.
- A further option is to create puzzles, poems or riddles for others to find.
- Themed caches are also recommended, possibly related to the site where the cache is hidden.

Geocachers are expected to replace whatever they take with something else, or to leave the find intact.

The learners now create their own geocaches, and decide on where to hide them in the town or city.

- The OpenCaching app has some helpful advice on how to choose the best locations.
- Each cache needs to be at least 160m away from another.

The learners should each log the GPS coordinates of their cache and compare these, to be as accurate as possible.

Once the location has been logged, the group needs to submit the cache description and coordinates via the OpenCaching website (accessible via the app, too).

The learners share their geocache coordinates with their classmates, and seek out each other's caches.

- They should leave comments in the physical logbook for each cache.
- They shouldn't remove or change any items in our activity.

If there are any public geocaches in your city or area, encourage the learners to look for them outside of class time, using the OpenCaching app.

- They log their finds, including comments, in the app 'Logbook' when they find a cache – which can be virtual or real.
- They will need to create a 'free user' account, to log their comments.

In a subsequent lesson, ask the learners to describe the caches they found – whether via the OpenCaching app or their classmates' caches – and to share any comments they logged.

The learners should 'follow' their own caches in the OpenCaching community:

- They monitor any comments that are made about their caches.
- They reply, if necessary.

Taking feedback on board – to improve the contents or location of a cache, based on user reviews – is also an important part of geocaching.

Going mobile

Although this activity suggests the learners go mobile using the OpenCaching app, you can run it just for a class or school, and without recourse to an app:

- The learners create their geocache boxes and hide them in the school grounds.
- The GPS coordinates need to be accurate: each cache needs to be at least 10–15 metres from the next.

The learners can then share the GPS coordinates of each cache with each other, or with other classes in the school. The contents of each cache can be purely language-focused: with review quizzes, and/or reading and listening material.

There is also a geocaching app and community called Munzee (*http://www.munzee.com/*) that places QR codes, rather than physical objects, in public places – this may be worth exploring with your learners.

Geocaching combines several useful digital skills – familiarity with geolocation and GPS coordinates – and participating in a responsible manner in a wider online community (of geocachers, in this case).

Augmented reality

Augmented reality is based on geolocation, but enables the real world to be enhanced – or 'augmented'.

To understand this, imagine the following:
- You are standing in front of a famous building like the Coliseum in Rome.
- You open an augmented reality app on your internet-enabled smartphone. The app automatically opens your camera.
- You hold up your device to look at the Coliseum – you will see the building through your smartphone camera lens and, at the same time, a little box of text will pop up on the screen, with some further information and a link to an online article about the Coliseum.
- You can click on the link to find out more about the building.

The app has done the following:
- Figured out that you are standing in front of the Coliseum with geolocation data.
- Provided you with an overlay of information about the Coliseum on your smartphone screen.

Augmented reality (AR) activities with learners can involve them not just accessing AR data in the real world via their smart devices, but actually *producing* AR data for others to read.

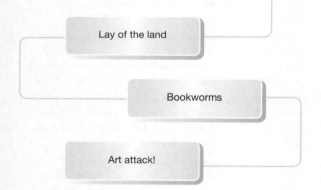

Lay of the land

Bookworms

Art attack!

Lay of the land
An AR brochure

In this extended activity, the learners research their town or city in greater depth, and prepare a multimedia, augmented reality advertising campaign brochure for the best places to visit.

Get the app
The learners need to install Wikitude and Layar on their mobile devices:
http://www.wikitude.com
http://www.layar.com
- Wikitude is an AR browser which allows users to look at the world around them through the camera on their connected device and receive information about the things surrounding them (places, sights, historical locations, restaurants – and more). This information is displayed as an 'information layer' on top of the camera picture.
- Layar allows users to access AR information via mobile devices in much the same way as Wikitude. It also enables them to create their own AR markers on the Layar website.

Here are two video tutorials to help you understand how each of these apps work:
- Wikitude: *http://goo.gl/Q2rkf5*
- Layar: *http://goo.gl/9rB0Ps*

Get ready
You register a 'developer' account at the Layar website and create a web page (called a 'campaign' in Layar) for the learners to add their resources to.

Ensure that the learners will have access during class to computers (not just mobile devices), to create AR markers and to upload their videos to the Layar campaign page you created.

Put the learners into six groups and give each group a theme connected to their town or city:
- Arts and events: *theatre, music, comedy, festivals, cultural events …*
- Architecture: *famous buildings, statues, monuments …*
- Sights: *zoos, museums, galleries …*
- Nature and outdoors: *beaches, forests, walking, sailing …*
- Accommodation: *hotels, guest houses, campsites …*
- Cafes and restaurants: *cafes, bars, pubs, takeaways …*

Ask each group to brainstorm suitable content for their subject area. If they are to recommend things to do or visit in their category:
- *What would they recommend, and why?*

Ask each group to make a 'Top 5' in their category.

Ask the learners to briefly share their thoughts with the rest of the class, and invite feedback or additions.

Get going

Now ask each group to open the Wikitude app. They visit the relevant section in the 'Categories' tab of Wikitude and explore the entries there.
- They will need an internet connection on their devices.
- They will not need to move around at this stage.

Ask them to review their top five choices, in the light of what Wikitude has shown them.
- *Would they make any changes?*

In their group, they should make any changes. Finally, for each category, each group chooses one entry from their top five choices, to include in the class AR brochure.

Regroup the six groups as three groups, combining two of the original groups in each case:
- Arts and events; Architecture
- Sights; Nature and outdoors
- Accommodation; Cafes and restaurants

Each group will now have two entries from the original, to work with.
- For each of their entries, they should take what they know, and what they can glean from Wikitude, to write an entry for the class AR brochure.
- This should be written, in the first instance, as a piece of text only.

The learners now visit the two places they have in their lists:
- They record a video of their text description of each of the two places.
- The video should include a view of the place (where possible), one of the learners speaking the text they have written, and the others saying why they like or recommend the place (or a brief personal anecdote to accompany it).
- They take a digital photograph of the two places.

Once they have done these two tasks, they return to class, to complete the next stages (or do them for homework). Ensure that each group has access to a connected computer which is logged in to your Layar developer account.
- The learners upload their final videos to YouTube (the easiest option).
- Alternatively, they send the videos to the computers they will be working on later (with this option, the videos will need to be converted to MP4 format if necessary).

Ask them to transfer the image they took to the computer they will be working from (eg via email).

You show them:
- How to add a page to the Layar campaign.
- How to upload the photo they have taken.
- How to link it to the video they uploaded to YouTube.

They can test the page by using their mobile devices and the Layar app, to check that their videos play.

Once the learners are happy with the layout of their page with video, they should 'publish' the page:

- Layar's website page provides help for all of these stages.
- Layar currently offers a free option for publishing a page with ads, or a pay option. The free version is sufficient for our purposes.

Provide individualised feedback on the videos, or provide feedback on common errors with the whole class.

Going mobile

This activity combines in-class research and discussion with a 'going mobile' stage in which the learners collect the data they need (in the form of video and photos) in situ.

If you work with young learners, you can ask them to create the same AR resources for Layar, but in this case:
- Ask the learners to make the video in the classroom or school – instead of at each venue – so that they don't need to leave the premises
- Show the learners how to find a Creative Commons image of each place (which they can download from the internet) so as to respect copyright.

Knowing how to use AR apps, and for what purposes, is another useful digital skill. Creating AR content is also an increasingly accessible skill, with relatively straightforward apps such as Layar providing the detailed tutorial support and simple interface necessary to achieve this.

||||||||

Bookworms
AR book reviews

Learners can use augmented reality (AR) apps, not only to find extra information about physical places or objects, but also to create their own AR information. In this activity, they use a program that creates AR 'markers' from real physical objects.

The learners create AR markers, and then 'read' each other's markers with their mobile devices. The markers in this activity are books, and the 'overlay' data are learner-created video book reviews.

Get the app
- The learners need to sign up for a 'developer' account in Layar (*https://www.layar.com/*).
- They also need to download the Layar app to their mobile devices.

Layar allows users to access AR information via mobile devices in much the same way as Wikitude (see *Lay of the land* on page 106). It also enables them to create their own AR markers on the Layar website.

Ensure that the learners will have access during class to computers – to create their markers and upload their videos to their Layar accounts. AR markers need to be created on a computer, not on a mobile device.

Get ready

Put the learners into pairs. Ask each pair to choose a favourite book – for example from the school library, or they can bring a favourite book from home.

Tell them to prepare a short book review about their chosen book. Put the following prompts on the board to help them structure their review:

- Title, author and publication date
- Summary of the story
- What they like about the book

Lower-level learners may want to script their reviews, while higher-level learners can work on their review in note form.

Help with language, and give plenty of time for them to rehearse their reviews before filming.

Get going

Ask each pair of learners to create one short video of their book review (two minutes maximum) on their mobile devices.

- If the learners are uncomfortable with appearing in the video, they can film the physical book and show different pages instead, while speaking.
- If they are not happy with the result at first, they can video-record and re-record until they are satisfied with the final version of their video.

Tell them to upload their final videos to YouTube (the easiest option) or to send the videos to the computers they will be working on later (with this option, the videos will need to be converted to MP4 format if necessary).

They should also take a photo of the book cover, and transfer the image to the computer they will be working from (eg via email).

The learners continue working in pairs, with one pair per computer, and log in to one of their Layar developer accounts.

- They follow the prompts in Layar to upload their book cover image, and add their video onto the image.
- They can test the page by using their mobile devices and the Layar app, to check that their videos play.

Once they are happy with the layout of their page with video, they should 'publish' the page.

- Layar's website page provides help for all of these stages.
- Layar currently offers a free option for publishing a page with ads, or a pay option. The free version is sufficient for our purposes.

Put the books at various points around the classroom or library.

- The learners open the Layar app on their mobile devices, and scan each book.
- Their devices must be connected to the internet – a video icon will appear on the book cover, only visible through the Layar app.

- They click on each video icon, and listen to each book review.

You could ask them to use headphones for this stage, and to work individually.

Listen to the reviews yourself, and take notes on common language errors or interesting uses of vocabulary or grammar, to share later.

Give the learners enough time to listen carefully to each book review. Encourage them to take notes for each review – for example, by filling in a grid like this:

	Book 1	Book 2	Book 3	Book 4
What is the title, author and publication date?				
What is the story about?				
What do the reviewers like?				
Would you read this book?				

Conduct class feedback about the reviews:

- *What kinds of books were reviewed?* (adventure, romance, thrillers, biographies, etc)
- *Which books sounded the most interesting?*
- *Which books would you read? Why?*
- *Which books wouldn't you read? Why?*

Finally, highlight and correct some of the common language errors with the class. You could also highlight any especially *good* examples of vocabulary or grammar from the reviews.

Going mobile

The books used for this activity can be in any language – as long as the book reviews are in English.

- The use of physical books as AR markers in this activity encourages the learners to go mobile by moving around the classroom or library.
- Creating and watching video book reviews on their mobile devices also makes for a rich multimedia experience.

This activity can also form part of a school-wide Book Fair:

- Several classes can participate and create book reviews.
- A Book Fair can be an effective way to draw the learners' attention to resources in the school library or resource centre.

The learners develop several advanced digital skills in this activity: multimedia skills in producing video, and creating AR markers via a developer account on an augmented reality website.

They can also share their work with a wider online audience if they choose to upload their markers to the Layar community site.

Art attack!
A multimedia art exhibition

The learners create an art gallery, featuring their favourite work of art, and a multimedia guide to the work itself.

Get the app

The learners need to install Aurasma:
www.aurasma.com

Aurasma is an augmented reality mobile service that allows you to create pairs of linked media called 'auras': these are called 'trigger images' and 'overlays', and they work together to provide a multimedia experience to a viewer. The basic idea is this:

- They open the Aurasma app on their devices.
- They point their device cameras at a trigger image.
- The trigger image is recognised by the Aurasma app.
- The app loads the associated overlay for them to view.

In practice, this means:

- A trigger image may be printed out and hung on a wall, or turned into a poster or some other visual display.
- Visitors can wander around viewing the images in Aurasma, which will automatically load the overlay.

In our activity, we are going to use photos of works of art for our trigger images and videos of learners talking about the works of art for our overlays, though the activity could be adapted for anything else that can be described on video.

Get ready

Ensure that everyone has installed the Aurasma app and created an account from the app itself. It will be helpful when explaining the project to have an example 'aura' ready, with a related video of yourself talking about a favourite work of art.

Ensure the learners will have access during class to printers, to print their chosen piece of art.

You can find out more about Aurasma here:
http://goo.gl/5RjuiK

Get going

Show the learners a photo (on screen or as a handout) of your favourite work of art:

- *Do they know what it is, and who painted it?*
- *What else do they know about it?*
- *What else would they like to know about it?*

Show them how Aurasma works:

- Open the app.
- Point it at your picture.
- Let the learners watch the video you recorded in advance.

Feed back as a whole group:

- *Were their questions answered by your video?*

If you have time and your learners have internet access, they can research any unanswered questions.

Ask each learner to think about a favourite piece of art, a picture or poster, a photograph – or something suitable in your context. They make notes on the following:

- *What is it called?*
- *Who created it?*
- *What do they know about the creator?*
- *What does it show/represent?*
- *What else did the creator do?*
- *Why do they like it?*

Show them how to create an aura using Aurasma – ensuring that they understand how to add the trigger image and video and save the overlay:

- They make their videos and add them to Aurasma.
- They print out and stick their trigger images (their pieces of art) to a wall, to create a classroom art gallery.
- They walk round.
- They choose the artwork they're interested in.
- They use the Aurasma app to learn more about it by watching the video (you could ask them to use headphones for this stage).

You then put them into pairs, to tell each other about the artwork they have just viewed.

Discuss as a whole class:

- *What were the favourite artworks?*
- *Who made the most compelling video?*

Going mobile

The mobile part of this activity is an extension in which the learners bring in 'works of art' from outside – these may be statues or pictures, or examples of graffiti or street art.

- They take a photo of their chosen artwork.
- They record their video about it 'in situ'.
- Their video should explain what the work of art is, where it can be found, why it was chosen …

Once they have their video and photo, they create a new aura in Aurasma. They share the location of the artwork on the class blog, wiki or by email.

Each learner should choose one of the new works of art:

- They find it in town.
- They use Aurasma to view the video.

Back in class, the learner reports back on the place they visited, what they found and what they found out about it.

There are other ways in which Aurasma might be used to encode useful information for language learners. You can experiment with video treasure hunts, vocabulary videos – or any number of uses where a creative combination of image and video might prove useful.

Going further ... in your institution

You and your learners may by now have worked with a range of mobile-based activities, and may even have experimented with some of the longer activities we have presented.

Excellent! You are now ready to broaden your approach, by helping your institution to develop a carefully designed mobile learning 'implementation plan'.

An institutional plan can help to support the *principled* implementation of device use. Also, when all of the institution is involved, the *success* of the initiative can more easily be measured and assessed over time, with the participation of a number of teachers and learners.

What follows is a ten-step plan for incorporating the use of mobile devices within your institution. Whether you are a teacher, teacher trainer, director of studies or manager, these steps will affect you *all* – and you will *all* need to be on board together, to carry out your plan effectively.

The ten steps are summarised below, and then each step of the plan is examined in detail.

1 Identify your reasons.
2 Assess your context.
3 Involve all the stakeholders.
4 Present your case.
5 Create learning plans.

6 Assign teacher champions.
7 Run a pilot phase.
8 Evaluate your pilot phase.
9 Extend the implementation plan.
10 Provide ongoing teacher development.

Step 1: Identify your reasons.

The first thing to be clear on is *why* your institution should want to develop a school-wide plan for the use of mobile devices. There are a number of possible reasons for this, and the reasons that most closely fit with your institutional context need to be clearly identified. Typical reasons – which may be pedagogical, promotional or economic – include:

- It will help teachers and learners develop what are key 21st-century skills and essential digital literacies.
- It will increase learners' motivation in class, and provide opportunities for extra out-of-class language learning.
- It will allow the learners to 'go mobile', and engage them in real-time communication in English – beyond the institution.
- It will increase the school's profile as a cutting-edge institution using the latest technologies, and provide a competitive edge.
- It will obviate the need for the school to invest in or maintain expensive hardware in large self-access centres or computer rooms for learners, especially if a BYOD ('Bring Your Own Device') approach is adopted.

In terms of real gains, the pedagogical reasons need to be put first. Although managers and directors may be attracted by the very real promotional and economic benefits, these will fade fast unless both teachers and learners are made aware of how the use of mobile devices is supporting classroom learning and also facilitating the use of the target language in the wider world.

> **Summary**
>
> Identify your reasons for implementing a mobile learning plan.
>
> Ensure that the learning benefits of your plan are foregrounded.
>
> Consider the benefits for all the stakeholders.

Step 2: Assess your context.

The next thing to take into consideration is context. Not just your institutional context – the wider socio-economic and educational contexts also need to be taken into account. To help you assess these and identify potential stumbling blocks, ask yourself the following:

Institutional context

- Does your institution have the necessary infrastructure, such as reliable wifi and IT staff, to support mobile learning?
- Is the physical space in the classrooms, and in the institution's buildings and grounds, conducive to having learners moving around with mobile devices?

Socio-economic context

- What economic investment is needed by your institution for implementation (including wifi, IT and hardware costs, and teacher training costs)?
- What devices do the students already have? Is there a gap between the 'haves' and the 'have-nots' within classes?
- What devices is the institution prepared to invest in?
- What system will be chosen? Will class sets be supplied? Or is BYOD the best approach? Or a hybrid BYOD/class set approach? And if the learners are encouraged to bring their own devices, will issues of personal safety (theft, mugging) arise?

Educational context

- How is the use of mobile devices perceived – in education in general? For example, do teachers, parents, and even learners, perceive them as potentially useful for learning, or as objects to be excluded from the classroom? If the latter, how will you change perceptions?
- Is the use of devices in your institution banned at the moment? What will be needed, to persuade administrators to overturn the ban? For example, will you be able to set up a meeting with management and other teachers, and present them with a well-argued case for allowing the use of devices in your school? See Step 4 below for how to do this.
- What sorts of learning activities do teachers and learners prefer to do in class? For example, do they believe that content delivery and drill-based grammar activities are the best way to learn English? If this is the case, how will you encourage them to use mobile devices for more communicative activities?
- Does your institution include teaching young learners? For example, how will issues such as e-safety, classroom management, or the (in)appropriate use of devices be dealt with?

Taking time to examine carefully how all these contexts – and their corresponding issues – will impact on your implementation plan can help you to identify potential challenges, and to think about how to deal with them.

Summary

Identify how infrastructure and physical space will affect the plan.

Consider how access to devices and other costs will affect the plan.

Think about how attitudes to devices, attitudes to learning activities and the age of your learners will affect the plan.

Step 3: Identify all the stakeholders.

It is important to think about who needs to be involved in your implementation plan: not just the main players – teachers and learners – but other players affected by the plan. If you are working with learners under 18, then their parents are also stakeholders, and their involvement is important too.

- Other stakeholders include: school directors or managers, who can provide support for the plan; IT staff, who will take care of the hardware and the technical support; and support staff such as librarians, who can deal with administering class sets of devices.

- Once you have identified the stakeholders, think about how you are going to get them on board. Exactly what are the roles and responsibilities each of them will have in implementing the plan?
- Once you have some initial ideas for this, informally canvas the various groups, and float the idea of a 'mobile implementation plan' for the school, as well as how each group might be involved.
- Once you have stakeholder input as to their perceptions of their roles and responsibilities, be very aware of their attitudes to the plan – both positive and negative – as you will need to address these in Step 4 below!

Developing a plan with stakeholder input from the beginning will ensure greater buy-in from everyone in the institution, and it will therefore enjoy a greater chance of success. Simply telling everybody what to do – with no process of consultation – will undermine your plan from the start.

Summary

Identify all the stakeholders.

Consider their roles and responsibilities in implementing the plan.

Gather input and opinions about the plan from the stakeholders.

Step 4: Present your case.

Now is the moment to convince the management of your institution of the value of a mobile learning implementation plan. You need to present your case and 'sell' the idea, especially if you work in an institution where the use of mobile devices is banned, or where concerns about young learners and mobile devices make the school administration (or the teachers) reluctant to use these.

- Ask management for permission to hold a formal meeting with the stakeholders, in order to present and argue the case for introducing the use of mobile devices in your school.
- Plan exactly what you are going to say, and how you are going to explain the benefits of learners having access to mobile devices. You can share some of the ideas outlined in Part A of *Going Mobile*, and also describe some of the activities from Part B. If possible, get some learner feedback on these activities to be able to share it.
- Ensure you address any negative reactions or concerns expressed by the stakeholders you canvassed in Step 3. Many typical concerns and challenges are outlined in Part A.
- Suggest inviting parents in for a special 'parents evening', in which the benefits of using mobile devices to support learning are clearly laid out, and examples from successful projects carried out by others are shared.
- Bear in mind that some groups of stakeholders may best be involved later on in the process – parents can always be persuaded of the benefits of mobile learning once the details of the plan are in place.

Essentially, presenting a persuasive case for implementing a mobile learning plan is about 'educating' stakeholders – especially management, other teachers, and parents of young learners – about the benefits that the use of mobile devices can bring to learning. If your stakeholders are receptive, you can move straight into Step 5. In some cases, however, management may want to review and discuss your idea privately, and come back to you at a later date with a decision on when and how to implement the plan.

Summary

Arrange a meeting with management and relevant stakeholders.

Plan how to present your case as persuasively as possible.

Address concerns from all the stakeholders.

Step 5: Create learning plans.

Once institutional support to move ahead with the plan has been gained, you can start to work in more detail on the specifics. First and foremost, work closely with other teachers to identify exactly how the use of devices can *enhance learning*. Most teachers will be working from a *syllabus*, whether imposed by the coursebook, the institution or the Ministry of Education, so it is important to integrate the activities within the framework of the syllabus. And identifying specific indicators to *measure success* is also part of developing a learning plan.

Enhancing learning

- Show the teachers example activities, such as those described in Part B, to help them create learning plans that include an effective use of mobile devices to improve communicative tasks.
- Show them how specific activities can achieve certain learning outcomes, and help them see how these activities can be enhanced through the use of mobile devices.
- Show them how learners can be slowly introduced to mobile-based activities through initially working with 'hands off' tasks, such as those described in Chapter One.
- Show them how mobile-based activities can become more varied over time, using text, images, audio and video, by demonstrating your favourite activities in Chapters Two to Five.
- Show how some activities can be carried out in the classroom, and that the concept of learners using their devices outside of class to 'go mobile' can also be introduced gradually.

Applying to the syllabus

- Demonstrate for the teachers how *you* integrate the activities into your own coursebook or syllabus.
- Encourage *them* to suggest some concrete ideas about how mobile devices can be used with *their* learners.
- Demonstrate *how* they can apply these ideas to their own coursebook or syllabus.
- Encourage them to look carefully at the first few units or sections, and to map specific mobile-based learning activities onto those:

  - They could ask their learners to contribute five facts about a coursebook topic to an online noticeboard. See *Sticky boards* on page 45.

  - They could ask their learners to create word clouds based on short coursebook texts or extracts. See *In the clouds* on page 46.

  - They could get their learners to collect photos for lexical sets related to the topic of a coursebook unit. See *Word bank* on page 60.

  - They could get their learners to use a voice search engine to research a topic related to a coursebook unit. See *Inventions* on page 73.

  - They could get their learners to draw and describe a process in the coursebook, using a screencasting app. See *Crazy sports* on page 80.

  - They could create stop-motion videos poems on a coursebook-related topic. See *Visual poems* on page 84.

Measuring success

- If you are asking the teachers to participate in mobile learning, having explained *why* you are introducing mobile learning in the first place (see Step 1), discuss *how* you are going to measure whether it is successful.
- If one of the reasons your institution is introducing mobile learning is to encourage learners to engage with communicative learning tasks outside the classroom – what evidence will you gather, to check that this is actually happening?
- If another of your reasons is to motivate groups of unmotivated teenage learners – how will you evaluate whether their level of motivation is actually increasing with the use of mobile devices?

In short, learning plans must have clear pedagogical aims for the activities, so you must show how they are related to the syllabus or coursebook, and address if and how the mobile-based activities will form part of assessment within all these frameworks.

Summary

Share mobile-based ideas and activities with the teachers.

Identify the learning outcomes for these, and relate them to the course syllabus.

Decide if, and how, mobile-based activities will be assessed.

Step 6: Assign teacher champions.

Once your plan is drawn up in some detail, you can turn your attention to teacher support. Identify one or two teachers to be your mobile learning plan 'champions'. These will be practising teachers who are interested in trying out using mobile devices with their own learners, and who are keen to be involved in the pilot phase (see Step 7 below).

■ If possible, champions should also be teacher trainers, so that they can share experiences and train their colleagues both during and after the pilot phase. However, the most important characteristic of a champion is that they are enthusiastic about the project, and are keen to learn. It is not necessary for the champion to be a technology expert – being a 'good teacher' is far more important than being knowledgeable about technology. Their enthusiasm for the mobile learning plan will inspire both their learners and the other teachers.

■ If possible, and if your school is using class sets, put the devices into the hands of these teachers as soon as possible, so that they can familiarise themselves with them in their own time, before starting to use them with their learners.

■ If possible, create an opportunity for some external training for staff at the outset – this could take the form of online training (eg an online course on the principles and practices involved in mobile learning). It could also involve external experts/trainers delivering face-to-face training sessions to teachers. This, of course, does not in any way obviate the implementation of crucial ongoing in-house teacher development. More on that in Step 10.

Given the importance of the pedagogical considerations we identified in Step 1, an emphasis on initial teacher training and development is fundamental to the success and uptake of any mobile learning plan, and is one area where institutions should not be 'economical'. Simply identifying a champion or two is *not* the end of the process.

Summary

Identify a few enthusiastic teacher champions.

Give these teachers access to devices and encourage them to explore.

Provide in-house and/or external training.

Step 7: Run a pilot phase.

With any new project, and especially with plans involving technology, it is always a good idea to start small. Unexpected challenges (and unexpected benefits) can arise, and it's easier to rectify a small-scale mistake than a large one that involves many learners and many teachers. So, start with a pilot phase – a 'pilot' is a short trial period, in which only a few people are going to take part.

■ Your pilot could involve just *one* teacher and one or two classes – depending on the size of your institution,

■ Your pilot could involve *several* teachers and a number of classes – of different levels.

In your pilot phase, context-specific challenges – such as an unexpected resistance from learners – can emerge, and will need to be dealt with. For example:

- If your learners are resistant to using mobile devices for communicative activities because they believe that drill-based grammar activities are the best way to learn, you can start by getting them to use mobile devices out of class with self-study grammar apps.
- If you start with activities suggested in Part B Chapter One, and then slowly introduce activities from Chapters Two to Five, you can gradually introduce more communicative in-class activities over time.

A pilot phase will allow you to rethink your plan, and deal with any unexpected challenges.

Summary

Start small.

Choose a few teachers and classes for the pilot phase.

Consider how the learning context and learner expectations might affect the pilot.

Step 8: Evaluate your pilot phase.

Involve the teachers and learners from the pilot in an evaluation, by soliciting feedback. Feedback from these stakeholders can then be used to improve the learning plans, and the project as a whole.
- Check how the *teachers* actually using the mobile devices feel about the experience, and whether they think it has been beneficial to their *teaching*.
- Check whether the *learners* found the integration of devices useful in their *learning*, and why/why not. For example, use a free online survey tool to gather anonymous feedback, and/or conduct personal interviews with individual learners.
- Remember to look at the *success* measurement indicators for learning (and any data gathered for this) that you developed during Step 5. For example, if encouraging the learners to use English out of class was specifically one of the aims of the plan, ensure you address this question when you collect feedback: Did they use English out of class? If so, did they feel this was beneficial? Why/why not?
- Ask for suggestions on how the pilot could be improved, and get feedback from *other* stakeholders for the evaluation. For example, a questionnaire could be sent to parents of young learners, to gather their impressions.

Collate the feedback from your pilot group of learners and teachers, and identify common themes or trends and useful suggestions for changes or improvements. You can now make any changes to your implementation plan, and to any of its stages, based on this feedback – in preparation for Step 9.

Summary

Collect feedback from the teachers and learners of your pilot via questionnaires or interviews.

Include questions related to the learning outcomes of the project.

Solicit suggestions for improving the project.

Step 9: Extend the implementation plan.

Now is the moment to roll out your implementation plan to a larger group of teachers and learners. You may want to involve all of the institution at this point, or to carry out a second, larger pilot phase. Once mobile devices are being used regularly in a larger pilot, or across the institution as a whole, it is still important to evaluate the plan regularly.
- You may want to change the way devices are accessed.
- You may want to switch from class sets to BYOD, or vice versa, depending on new developments in mobile technology, or on access to devices in your context.

- You may want to change the roles and responsibilities of the teachers in the plan.
- You may want to involve more (or fewer) champions, or provide more (or fewer) teacher development opportunities, depending on uptake or resistance among the teachers.

The important thing to remember is that your plan should not be set in stone. It needs to be flexible so it responds to the needs of all the stakeholders. It also needs to respond to pedagogical and technological developments in the field over time. Reviewing your mobile learning plan, and evaluating its efficacy, is an ongoing developmental process.

Summary

Extend the plan to a larger group of teachers and learners.

Decide when and how to evaluate the plan regularly.

Review the plan, and make periodic changes or adjustments as necessary.

Step 10: Provide ongoing teacher development.

Now that your plan is being implemented more broadly across your institution as a whole, don't forget the teachers! Plan to hold regular teacher development sessions, to ensure that the teachers discuss and share best practice about how they are using mobile devices with their learners.

This also provides them with a space to explore any challenges that may arise, and to find ways to overcome them. Here is one way to provide ongoing development and support for your teachers in a mobile learning project:

- Every week, hold short teacher development sessions or 'coffee mornings' (eg for just 30 or 45 minutes) in which teachers come together and share their ideas and experiences.
- At the end of every session, suggest one short, achievable, mobile-based activity for the teachers to try out with their learners during the upcoming week. You could choose activities from Part B of *Going Mobile* – starting from Chapter One, for teachers who are new to teaching and learning with mobile devices.
- In the following week's session, share experiences of how the chosen activity worked with learners, and pool any learning points, or 'things to keep in mind' for the next time they do the activity.
- Then set a new achievable task.

Some schools start projects involving new technologies by initially working only with those teachers who are keen to join in, rather than obliging an entire – possibly unwilling – staff of teachers to join.

- This has the advantage of ensuring that implementation is likely to be successful, and that a great deal of enthusiasm is generated around the project.
- This soon spreads among teachers and learners, and initially reluctant teachers can become intrigued and interested to learn more, and may eventually want to join in.

By ensuring that teacher development sessions such as the one described are voluntary – but ongoing and cyclical (repeated from the beginning, about once a term) – other teachers don't miss out, and can decide to join in your mobile learning project at a later stage.

Summary

Hold regular teacher development sessions to share ideas and best practice.

Set achievable tasks for teachers to try with their classes as part of their development.

Ensure that involvement in the project is voluntary and supported.

Looking back ...

Looking back over the activities you have carried out with your classes,
you ought to see the benefits that using the unique affordances of smart mobile devices – QR code readers,
tactile screens, geolocation, augmented reality – can bring to tasks, by making them truly 'mobile'.

To conclude:

1 **Learning** can be taken out of the classroom – as learners are encouraged to integrate information from their surroundings into classroom work.

2 Teachers can give **language input** in these 'going mobile' activities at the point of need – for example, while creating a final digital product – so the focus is primarily on communication.

3 Learners can create rich **multimedia products** – integrating text, image, audio and/or video – based on the information gathered outside the classroom.

4 'Going mobile' tasks can be introduced gradually – by starting with the technically **less demanding** (QR codes) and moving to the **more challenging** (creating augmented reality markers).

5 Longer-term **project work** can be incorporated over several lessons – which can then form part of assessment.

Looking back over the steps you have followed within your institution,
you ought to see the benefits that having a well-thought-out implementation plan can bring you.
But not only *you* – your colleagues, your learners and your institution at large.

To conclude:

1 Institutional support will ensure that **basic infrastructure**, such as reliable wifi or access to extra devices, is available.

2 Teachers and learners will understand **why and how** mobile devices can support their language learning.

3 Teachers and learners will be clear about **when and where** mobile devices can be used, and young learners will have guidelines about appropriate and inappropriate use.

4 Teachers will receive ongoing **teacher development and support** in the best use of devices, will be able to share best practice and activity ideas, and will find solutions to challenges together.

5 Learners will use their devices for learning with different teachers in different classes across the institution, leading to a smoother **learning experience**, rather than with some teachers allowing device use and others not.

Looking forward ...

But this is only the beginning.

You will be learning to deal successfully with the challenges we have presented in this book – and many more! – as you start going mobile.

And you will be continuing to go forward as an educator who is constantly developing those necessary digital literacies in the fast-moving 21st century.

From the editors

Going Mobile is a book that couldn't have been written several years go. Gavin is reminded by his personal collection of gadgets how quickly the world has changed over the past years. And Nicky points to the central role mobile devices have now come to play in her life, in her teaching and in her training.

Going Mobile aims to fill the gap they both felt still exists in the support offered to teachers in how to use these devices in the English Language Teaching classroom – in a principled and pedagogically sound way – and it succeeds admirably.

 **Part A** introduces us to teaching with hand-held devices, and the apps we can use. It covers the important questions we often ask, the issues we frequently face and the challenges we invariably encounter. And provides all the answers and advice that the authors felt was missing.

 **Part B** is full of activities – from the first chapter which takes an analytical yet entertaining look at the use of mobile devices, to the other four chapters where the authors approach, in a staged and sensible manner, the exploitation of text, image and audio, before pulling everything together to work with video.

 **Part C** takes us further, both in our classes, by way of longer activities and projects, and in our institutions, where we are encouraged to incorporate hand-held learning and teaching progressively and systematically – again, in a staged and sensible manner.

Going Mobile is very much a child of the 21st century – where, as teachers or other people involved in language education, we are definitely invited to come of age.

Mike Burghall
Lindsay Clandfield

From the publisher

DELTA TEACHER DEVELOPMENT SERIES

A pioneering award-winning series of books for English Language Teachers
with professional development in mind.

**Storytelling
With Our Students**
by David Heathfield
ISBN 978-1-905085-87-3

The Autonomy Approach
by Brian Morrison and
Diego Navarro
ISBN 978-1-909783-05-8

Spotlight on Learning Styles
by Marjorie Rosenberg
ISBN 978-1-905085-71-2

The Book of Pronunciation
by Jonathan Marks and
Tim Bowen
ISBN 978-1-905085-70-5

The Company Words Keep
by Paul Davis and
Hanna Kryszewska
ISBN 978-1-905085-20-0

Digital Play
by Kyle Mawer and
Graham Stanley
ISBN 978-1-905085-55-2

Teaching Online
by Nicky Hockly with
Lindsay Clandfield
ISBN 978-1-905085-35-4

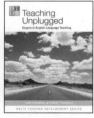

Teaching Unplugged
by Luke Meddings and
Scott Thornbury
ISBN 978-1-905085-19-4

Culture in our Classrooms
by Gill Johnson and
Mario Rinvolucri
ISBN 978-1-905085-21-7

The Developing Teacher
by Duncan Foord
ISBN 978-1-905085-22-4

Being Creative
by Chaz Pugliese
ISBN 978-1-905085-33-0

The Business English Teacher
by Debbie Barton,
Jennifer Burkart and
Caireen Sever
ISBN 978-1-905085-34-7

For details of these and future titles in the series, please contact the publisher: *E-mail* info@deltapublishing.co.uk
Or visit the DTDS website at www.deltapublishing.co.uk/titles/methodology